# Preserving the Legacy

# Communication Skills for the Environmental Technician

HMTRI

JOHN WILEY & SONS, INC.

New York / Chichester / Weinheim / Brisbane / Singapore / Toronto

Technical Illustration: Richard J. Washichek, Graphic Dimensions, Inc.
Research: Judy Sullivan

This book is printed on acid-free paper. ♾

This material is based on work supported by the National Science Foundation under Grant No. DUE94-54521. Any opinions, findings, and conclusions or recommendations expressed in this material are those of the author(s) and do not necessarily reflect those of the National Science Foundation.





***Library of Congress Cataloging-in-Publication Data:***

ISBN: 0471 29981-2

10 9 8 7 6 5 4 3 2 1

# Table of Contents

# Preface

*Communication Skills for the Environmental Technician* is designed to provide individuals studying environmental technology with a resource that will not only enhance their communication skills, but also their chances of getting a job. It is the sixth volume in the ***Preserving the Legacy*** series developed by INTELECOM Intelligent Telecommunications in association with the Partnership for Environmental Technology Education (PETE).

An underlying theme throughout this book is the close correlation between communication skills and employability. In fact, in a survey conducted by the Center for Occupational Research & Development (CORD), industry leaders ranked communication skills as even more important than computer and technical skills. The purpose of this book, therefore, is to provide the principles underlying good communication skills, models, activities, and strategies that can be used during and after the job search process. Although frequent references are made to the use of computers and the availability of resources on the World Wide Web, they are not required for the use of this book.

Many teachers in the environmental technology field are themselves working professionals, aware of the need to improve their students' communication skills. In the past, there have been two hurdles: finding the time within the framework of existing courses, and the availability of an appropriate resource. This book has been designed to address both of these problems. First, related skills are subdivided into eight chapters. Learning objectives are given at the beginning of each chapter to guide student progress. The chapter is then subdivided into sections, where the discussion and models provided are designed to help students achieve these objectives. At the end of each section, one or more activities are provided for students to practice and demonstrate their newly acquired skills. Students are expected to accomplish most of this work independently, therefore, requiring little valuable instructor or class time. Second, this book is designed for use in conjunction with the other textbooks in the ***Preserving the Legacy*** series.

By using the Activities Table after this Preface, teachers can quickly identify activities that emphasize skill types appropriate for use with each of the textbooks in the ***Legacy*** series. The particular type of skill emphasized in each of the activities is not only identified, but its degree of difficulty is ranked as either basic or advanced. At the same time, the book is equally applicable for those students using other textbooks or the independent learner who simply wants to improve their chances of getting a job and performing it well.

The sequence of chapters in this book is not intended to suggest an absolute order of use. It is suggested, however, that Chapters 1 and 2 be assigned early in the environmental technology program. The primary intent of Chapter 1 is to firmly establish the connection between all forms of communication skills and employability. The Secretary's Commission on Achieving Necessary Skills (SCANS) assessment instrument included in this chapter provides the student with a tool to evaluate their own strengths and weaknesses and invites them to set goals early in their program of study.

Foundational reading and writing skills are introduced in Chapter 2, including outlining skills and grammar awareness. Outlining is a valuable tool when reading difficult documents and preparing to write or speak. Chapters 3, 4, and 5 are designed to improve the student's ability to prepare written documents that are frequently the responsibility of the environmental technician. In Chapter 3 the emphasis is on writing many different kinds of business letters and memos. In Chapter 4, the emphasis shifts to written plans that regulations often require. Chapter 5 provides help in locating and completing the myriad of required reporting forms. The Internet is recommended as a resource for obtaining up-to-date forms from the various governmental agencies.

Starting with Chapter 6, the emphasis switches from reading and writing skills to improving oral communication skills. The focus of Chapter 7 is on finding and landing a job. This chapter contains information on developing a portfolio and résumé, another topic that should be introduced early in the student's program of study. The final chapter considers various interpersonal skills and includes several activities to make students more aware of the importance of teamwork.

Many of the examples used in this book are from real companies or are adapted from real events. Included as Appendices are two sections taken from the Code of Federal Regulations (CFRs) and two Material Safety Data Sheets (MSDS) that are intended for use in several of the activities. Two fictitious scenarios provide background information for a number of different activities. A much more complicated EPA training scenario is included as Appendix 5 It is provided for students who have considerable experience in the field or are nearing the end of their environmental technology program.

## Acknowledgments

The authoring team wishes to acknowledge the organizations that have made this project possible:

- INTELECOM, Intelligent Telecommunications and the National Science Foundation (NSF) for making this project possible.
- The Partnership for Environmental Technology Education (PETE) who nominated the members to serve on the National Academic Council (NAC).
- The NAC members for their reviews and helpful suggestions during the development of the manuscript: Ann Boyce, Bakersfield College; Doug Feil, Kirkwood Community College; A.J. Silva, Eastern Idaho Technical College; Dave Boon, Front Range Community College; Bill Engel, University of Florida; Eldon Enger, Delta College; Doug Nelson, SUNY Morrisville; Stephen Onstot, Esq., Burk, Williams, and Sorensen; and Ray Seitz, Columbia Basin College.
- Graphic Dimensions for their talent in converting the manuscript into an attractive and useful text.
- Judy Sullivan for keeping us communicating and handling the many project details.
- Howard Guyer, Academic Team Leader and Ann Boyce, member of the NAC, for their efforts to shape the final manuscript.
- Chris Robinson, INTELECOM Editor, for her expertise in the overall organization of the manuscript, rewriting suggestions, and attention to detail.

The team also heartily thanks the following individuals and companies for contributing documents and advice: Pat Berntsen, HMTRI; Bette Beshears, Lamar University – Port Arthur; David Foss, Ph.D., Rock Island Arsenal; Doug Feil, HMTRI; Steve Fenton, Highline Community College; John Gaines, Forest Road Consulting, Inc.; Sally Gaines, Eastern Iowa Community College District; Ellen Kabat, Ph.D., Eastern Iowa Community College District; John Konefes, Iowa Waste Reduction Center; Cynthia Lake Cary, HMTRI; Steve MacKenzie, Central Florida Community College; Rick Peters, Monsanto; Brian Stone, IEFM Consulting Engineers; and Cheryl Stith, Cosumnes River Community College.

Diane Dickens Gere, Instructional Designer
HMTRI
ATEEC, 1999

## The HMTRI Writing Team

The Hazardous Materials Training and Research Institute or HMTRI, which is listed as the author of this book, was established in 1987 by the Eastern Iowa Community College District, headquartered in Davenport, Iowa, and Kirkwood Community College in Cedar Rapids, Iowa. The purpose of the Institute is to promote worker protection and the maintenance of a clean and safe environment through education and training. HMTRI is recognized as a national center of excellence by several federal agencies including the U.S. EPA, the National Science Foundation, and the National Institute of Environmental Health Sciences.

**Diane Gere**
Diane Gere is the coordinator and HMTRI editor for this textbook. Beginning her instructional design work for HMTRI in 1989, Diane Gere works on curriculum projects associated with the Advanced Technology Environmental Education Center (ATEEC), which is funded by the National Science Foundation. Diane has over 20 years of experience as an educator, including seven as a community college teacher of composition, editing, communication skills, and speech. Besides leading the writing team for this book, she led the writing teams for INTELECOM's *Site Characterization: Sampling and Analysis*; HMTRI's *Hazardous Materials Industrial Processes*, *EPA Regulations II*, and *Hazardous Materials Sampling and Monitoring*; and the Iowa Waste Reduction Center's *Solvent Waste Reduction and Recycling – Practical Advice for Small Business.*

**Melonee Docherty**
Having joined HMTRI in 1995 as an instructional designer for environmental compliance and technology textbook revisions, Melonee Docherty also works on projects for ATEEC and industry clients. Melonee has nearly 20 years of experience as a technical writer and editor, including ten years spent coordinating and managing technical publications teams and projects.

**Lea Campbell**
Lea Campbell is the Executive Director for the South Central Partnership for Environmental Technology Education (SC PETE). Prior to assuming the directorship of SC PETE, Ms. Campbell served as Department Head of Industrial Technology at Lamar University Port Arthur, TX. Ms. Campbell holds both a BS and MA in mathematics from Eastern New Mexico University and has published several undergraduate mathematics texts.

**ACTIVITIES TABLE**
**By Skill Type and Usage with Other *Preserving the Legacy* Textbooks**

| Chapter | Section | Activity | Other *Preserving the Legacy* Series Books | | | | | Level of Difficulty | |
|---|---|---|---|---|---|---|---|---|---|
| | | | Introduction to Environmental Technology | Site Characterization | Basics of Toxicology | Basics of Industrial Hygiene | Ind. Processes & Waste Stream Management | Basic | Advanced |
| 1 | 1-1 | A | R/W/O | R/W/O | R/W/O | R/W/O | R/W/O | R/W/O | |
| 1 | 1-1 | B | R/W/O | R/W/O | R/W/O | R/W/O | R/W/O | R/W/O | |
| 1 | 1-2 | A | R/W | | R/W | R/W | | R/W | |
| 1 | 1-2 | B | R/W | | R/W | | | R/W | |
| 1 | 1-3 | A | W | | W | W | W | W | |
| 1 | 1-3 | B | W/O | W/O | W/O | W/O | W/O | W/O | |
| 1 | 1-3 | C | R/W | | R/W | R/W | R/W | R/W | |
| 1 | 1-4 | A | R | R | R | R | R | R | |
| 1 | 1-4 | B | R/W/L | R/W/L | R/W/L | R/W/L | R/W/L | R/W/L | |
| 1 | 1-4 | C | O | | | | | O | |
| 1 | 1-4 | D | R/W | | R/W | R/W | R/W | R/W | |
| 2 | 2-2 | A | R | | R | R | R | R | |
| 2 | 2-2 | B | R/W | | R/W | R/W | R/W | | R/W |
| 2 | 2-3 | A | W/O/L | W/O/L | W/O/L | W/O/L | W/O/L | | W/O/L |
| 2 | 2-3 | B | | R/W | R/W | R/W | R/W | | R/W |
| 2 | 2-4 | A | R/W | | | | | R/W | |
| 2 | 2-4 | B | R/W | R/W | R/W | R/W | R/W | | R/W |
| 2 | 2-4 | C | R/W | R/W | R/W | R/W | R/W | | R/W |
| 2 | 2-4 | D | R/W | | R/W | R/W | | | R/W |
| 2 | 2-5 | A | I/R/W | I/R/W | I/R/W | I/R/W | I/R/W | | I/R/W |
| 2 | 2-5 | B | I/R/W | | I/R/W | I/R/W | I/R/W | | I/R/W |
| 2 | 2-6 | A | R/W | R/W | R/W | R/W | R/W | | R/W |
| 2 | 2-6 | B | R/W | R/W | R/W | R/W | R/W | | R/W |
| 2 | 2-6 | C | R/W | | R/W | R/W | R/W | R/W | |
| 3 | 2-7 | A | R/W | | R/W | R/W | R/W | R/W | |
| 3 | 2-7 | B | D | D | | D | D | D | |
| 2 | 2-8 | A | R/W | R/W | | | R/W | R/W | |
| 2 | 2-8 | B | R/W/T | R/W/T | | | R/W/T | R/W/T | |
| 3 | 3-1 | A | R | R | R | R | R | R | |
| 3 | 3-1 | B | R/W | R/W | R/W | R/W | R/W | R/W | |
| 3 | 3-2 | A | R/W | | | R/W | R/W | R/W | |

| Chapter | Section | Activity | Other *Preserving the Legacy* Series Books | | | | | Level of Difficulty | |
|---|---|---|---|---|---|---|---|---|---|
| | | | Introduction to Environmental Technology | Site Characterization | Basics of Toxicology | Basics of Industrial Hygiene | Ind. Processes & Waste Stream Management | Basic | Advanced |
| 3 | 3-2 | B | R/W | R/W | | | | R/W | |
| 3 | 3-2 | C | R/W | | | R/W | R/W | R/W | |
| 3 | 3-2 | D | R/W | R/W | | R/W | | R/W | |
| 3 | 3-2 | E | R/W | R/W | R/W | R/W | R/W | R/W | |
| 3 | 3-3 | A | R/W | R/W | R/W | R/W | R/W | R/W | |
| 3 | 3-3 | B | R/W | | R/W | R/W | R/W | R/W | |
| 3 | 3-3 | C | R/W | R/W | R/W | R/W | R/W | R/W | |
| 3 | 3-3 | D | R/W | R/W | R/W | R/W | R/W | R/W | |
| 3 | 3-3 | E | R/W | R/W | R/W | R/W | R/W | R/W | |
| 3 | 3-3 | F | R/W | R/W | R/W | R/W | R/W | R/W | |
| 3 | 3-3 | G | R/W | R/W | | | | R/W | |
| 3 | 3-3 | H | R/W | R/W | R/W | R/W | R/W | | R/W |
| 3 | 3-3 | I | R/W | R/W | R/W | R/W | R/W | | R/W |
| 3 | 3-3 | J | R/W | R/W | | | | R/W | |
| 3 | 3-3 | K | R/W | R/W | | | | | R/W |
| 3 | 3-3 | L | R/W | R/W | | | R/W | | R/W |
| 3 | 3-3 | M | | R/W | R/W | R/W | R/W | | R/W |
| 3 | 3-3 | N | R/W | R/W | | | R/W | | R/W |
| 3 | 3-4 | A | R/W | | | R/W | R/W | R/W | |
| 3 | 3-4 | B | | | | R/W | R/W | | R/W |
| 3 | 3-4 | C | R/W | R/W | | | R/W | R/W | |
| 3 | 3-4 | D | R/W/C | R/W/C | R/W/C | R/W/C | R/W/C | | R/W/C |
| 3 | 3-4 | E | | | | R/W | R/W | | R/W |
| 4 | 4-1 | A | L/W/O | L/W/O | L/W/O | L/W/O | L/W/O | L/W/O | |
| 4 | 4-1 | B | R/W | R/W | R/W | R/W | R/W | R/W | |
| 4 | 4-2 | A | R/W | | | | R/W | | R/W |
| 4 | 4-2 | B | R/W | | | | R/W | | R/W |
| 4 | 4-2 | C | R/W | R/W | R/W | R/W | R/W | | R/W |
| 4 | 4-2 | D | R/W | R/W | R/W | R/W | R/W | | R/W |
| 4 | 4-3 | A | R/W | R/W | R/W | R/W | R/W | R/W | |
| 4 | 4-3 | B | R/W | R/W | R/W | R/W | R/W | | R/W |
| 4 | 4-3 | C | R/W | R/W | R/W | R/W | R/W | | R/W |
| 4 | 4-4 | A | R/W | | | R/W | R/W | R/W | |

| Chapter | Section | Activity | Other *Preserving the Legacy* Series Books | | | | | Level of Difficulty | |
|---|---|---|---|---|---|---|---|---|---|
| | | | Introduction to Environmental Technology | Site Characterization | Basics of Toxicology | Basics of Industrial Hygiene | Ind. Processes & Waste Stream Management | Basic | Advanced |
| 4 | 4-4 | B | R/W | R/W | R/W | R/W | R/W | | R/W |
| 4 | 4-4 | C | R/W | R/W | R/W | R/W | R/W | | R/W |
| 4 | 4-4 | D | R/W/T | R/W/T | R/W/T | R/W/T | R/W/T | | R/W/T |
| 4 | 4-4 | E | R/W | R/W | R/W | R/W | R/W | | R/W |
| 4 | 4-5 | A | W/D | W/D | W/D | W/D | W/D | W/D | |
| 5 | 5-1 | A | | R/W/T | | R/W/T | R/W/T | | R/W/T |
| 5 | 5-1 | B | | R | | R | R | R | |
| 5 | 5-1 | C | R/W | R/W | R/W | R/W | R/W | R/W | |
| 5 | 5-2 | A | | | R/W | R/W | R/W | R/W | |
| 5 | 5-2 | B | R/W/C | R/W/C | | R/W/C | R/W/C | | R/W/C |
| 6 | 6-1 | A | L/W | L/W | L/W | L/W | L/W | L/W | |
| 6 | 6-1 | B | L/W/O/T | L/W/O/T | L/W/O/T | L/W/O/T | L/W/O/T | L/W/O/T | |
| 6 | 6-2 | A | L/W/O | L/W/O | L/W/O | L/W/O | L/W/O | L/W/O | |
| 6 | 6-2 | B | W | W | W | W | W | W | |
| 6 | 6-3 | A | W | | | W | W | W | |
| 6 | 6-3 | B | W/D | | | W/D | W/D | | W/D |
| 6 | 6-4 | A | | | | W/O/D | W/O/D | | W/O/D |
| 6 | 6-4 | B | R/W/O | | | R/W/O | R/W/O | | R/W/O |
| 6 | 6-5 | A | R/W | R/W | R/W | R/W | R/W | R/W | |
| 6 | 6-5 | B | L/W/D | L/W/D | L/W/D | L/W/D | L/W/D | | L/W/D |
| 6 | 6-5 | C | R/W/I | R/W/I | R/W/I | R/W/I | R/W/I | | R/W/I |
| 7 | 7-1 | A | R/W/I/O | R/W/I/O | R/W/I/O | RW/I/O | RW/I/O | R/W/I/O | |
| 7 | 7-1 | B | R/W/I | R/W/I | R/W/I | R/W/I | R/W/I | R/W/I | |
| 7 | 7-1 | C | L/W/O | L/W/O | L/W/O | L/W/O | L/W/O | L/W/O | |
| 7 | 7-2 | A | R/W | R/W | R/W | R/W | R/W | R/W | |
| 7 | 7-2 | B | R/W | R/W | R/W | R/W | R/W | | R/W |
| 7 | 7-3 | A | R/W/O | R/W/O | R/W/O | R/W/O | R/W/O | R/W/O | |
| 7 | 7-3 | B | R/W | R/W | R/W | R/W | R/W | | R/W |
| 7 | 7-3 | C | R/W | R/W | R/W | R/W | R/W | R/W | |
| 7 | 7-3 | D | R/W | R/W | R/W | R/W | R/W | | R/W |
| 7 | 7-3 | E | R/W | R/W | R/W | R/W | R/W | | R/W |
| 7 | 7-4 | A | R/W | R/W | R/W | R/W | R/W | R/W | |
| 7 | 7-4 | B | | L/W/O | L/W/O | L/W/O | L/W/O | | L/W/O |

| Chapter | Section | Activity | Other *Preserving the Legacy* Series Books | | | | | Level of Difficulty | |
|---|---|---|---|---|---|---|---|---|---|
| | | | Introduction to Environmental Technology | Site Characterization | Basics of Toxicology | Basics of Industrial Hygiene | Ind. Processes & Waste Stream Management | Basic | Advanced |
| 7 | 7-4 | C | | W | W | W | W | W | |
| 7 | 7-4 | D | | R/W | R/W | R/W | R/W | | R/W |
| 7 | 7-4 | E | R/W | R/W | R/W | R/W | R/W | R/W | |
| 7 | 7-4 | F | | W/T | W/T | W/T | W/T | | W/T |
| 8 | 8-1 | A | L/W | L/W | L/W | L/W | L/W | L/W | |
| 8 | 8-1 | B | W/T/O | W/T/O | W/T/O | W/T/O | W/T/O | | W//T/O |
| 8 | 8-1 | C | W/T/O | W/T/O | W/T/O | W/T/O | W/T/O | | W/T/O |
| 8 | 8-2 | A | L/W | L/W | L/W | L/W | L/W | L/W | |
| 8 | 8-2 | B | | L/O/T | L/O/T | L/O/T | L/O/T | | L/O/T |
| 8 | 8-2 | C | | L/O/T | L/O/T | L/O/T | L/O/T | | L/O/T |
| 8 | 8-3 | A | T/W | T/W | T/W | T/W | T/W | | T/W |
| 8 | 8-3 | B | T/W | T/W | T/W | T/W | T/W | T/W | |
| 8 | 8-3 | C | T/O/W/D | T/O/W/D | T/O/W/D | T/O/W/D | T/O/W/D | | T/O/W/D |
| 8 | 8-4 | A | T/W | T/W | T/W | T/W | T/W | T/W | |
| 8 | 8-4 | B | | | | T/W/O | T/W/O | | T/W/O |
| Appendix 5 | | A-1 | | R/W | R/W | R/W | R/W | | R/W |
| Appendix 5 | | A-2 | | R/W | R/W | R/W | | | R/W |
| Appendix 5 | | A-3 | | R/W | | | | | R/W |
| Appendix 5 | | A-4 | | R/W | R/W | | | | R/W |
| Appendix 5 | | B | | R/W | R/W | | | | R/W |
| Appendix 5 | | C | | R/W/O | R/W/O | R/W/O | | | R/W/O |
| Appendix 5 | | D | | R/D/O | R/D/O | R/D/O | | | R/D/O |
| Appendix 5 | | E | | R/W/T | R/W/T | R/W/T | | | R/W/T |

| Skills Legend | |
|---|---|
| C = Computer | R = Reading |
| D = Drawing | T = Team Building |
| L = Listening | I = Internet |
| O = Oral | W = Writing |

1

# Communication Skills Overview

## Chapter Objectives

Upon completing this chapter, the student will be able to:

1. **Explain** the importance of communication skills in the workplace.
2. **Describe** effective communication.
3. **Identify** the elements necessary for effective communication.
4. **Evaluate** and determine if communication is appropriate for a specific audience.

## Chapter Sections

# 1-1 How Communication Skills Affect Employability

## Concepts

- Communications skills are as valuable as technical expertise.
- Communication skills help employees to become better at job-related skills.

Visualize a fictitious company. Ms. Dennis, an engineer, submits a hiring requisition to personnel for an environmental technician. Ms. Brown, the Director of Human Resources, is responsible for filling the position.

Ms. Brown begins by sorting through 30 resumes. Half of the applicants are quickly eliminated because their cover letters are so poorly written it is apparent that they do not have the writing skills necessary for the position. From the 15 resumes that are left, four candidates failed to follow the directions stated in the classified ad. One person sent a resume that did not include current job information. Another candidate failed to provide a list of references.

Having narrowed the field to nine applicants, Ms. Brown next attempts to conduct telephone interviews with each of the remaining candidates. The telephone number on one resume results in a "no longer in service" message, so Ms. Brown tosses it into the discard pile. Four of the applicants are unprofessional during the telephone interview. The four remaining candidates are invited to schedule interviews.

During the face-to-face meetings, Ms. Brown and Ms. Dennis notice a distinct lack of communication skills in two of the interviewees. One does not make eye contact while speaking, and the other cannot be induced to elaborate on his responses beyond a word or two.

The task of hiring the best candidate gets tougher. Each of the two final candidates seems capable of doing the job. They meet the education qualifications, have organized resumes and cover letters, and they are both polite and personable. In the final analysis, Candidate A has the advantage – she received an excellent recommendation for her communication skills and expertise at working with others. The interviewers saw these skills as a particular asset because the company often uses a team approach. Candidate A gets the job.

The situation described above is less complicated than most hiring processes, but it demonstrates the high value businesses place on communication skills.

> "Instructors get so locked into teaching students technical proficiency in the field of study that we sometimes overlook which skills employers really desire."
>
> Steven MacKenzie,
> Central Florida College

Talk to a Human Resources manager about the value of an employee who can communicate effectively with coworkers and clients, and work cooperatively in a team setting. You'll find that these characteristics are often considered equal to technical expertise, particularly in an entry-level position.

> "Students who are flexible, innovative, open-minded, and computer literate, and who have a desire to learn, will get the job because they will help the employer succeed. 'Pay for performance' is the wave of the future; students must bring to the table employability skills if they want to be successful in the workplace.
>
> "About two years ago a student wrote me a thank you note on a course evaluation. He told me his wife had been trying to get him to use the computer for a while but he was afraid to try. He said my class was the first one that exposed him to the computer constantly. Although I did not have a computer in the room, in almost every lecture I used information gathered from the Internet to answer questions, challenge students, show real world applications, and make it fun. He realized that he didn't have anything to be afraid of, and now his wife has to jockey for time on the computer. This student learned that by being flexible and willing to learn, he could expect more of himself. When we challenge ourselves, the sky is the limit!"
>
> Cheryl Stith, Consumnes River
> Community College, California

In 1990, the U.S. Secretary of Labor appointed leaders from education, the business community, labor unions, and health and social service organizations to the Secretary's Commission on Achieving Necessary Skills (SCANS). The Commission's study yielded a list of "Foundation Skills" and "Workplace Competencies" needed for employment. (A copy of the SCANS list has been included in Activity A at the end of this section, and you are invited to do a self-assessment.)

The Foundation Skills include basic skills (reading, writing, and arithmetic), thinking skills, and personal qualities. The Workplace Competencies include the abilities to manage resources, use information, work with others, understand systems, and use technology. All of the skills and competencies in the SCANS list are highly desired by employers of environmental technicians (see Figure 1-1) and various industry groups (see Figure 1-2).

Driving all of the above-mentioned skills and competencies is the ability to communicate, which affects nearly every aspect of job performance. For example, employers want responsible workers who will stay with the job until it is done. An employee who cannot get other workers to understand his ideas is more likely to become frustrated and quit than someone who can express his thoughts clearly. The employee who communicates effectively is apt to work better within a team, get along with others, and consequently finish the work with less stress and frustration.

Your communication skills will make the difference between getting or not getting a job and between keeping the job or losing it.

*The National Voluntary Skills Standard in Hazardous Materials Management Technology* (HMMT) lists technical skills, related academic skills, and employability skills needed to perform tasks in the environmental technology occupation. It was developed by a team of experts in environmental technology for the Center for Occupational Research & Development (CORD). According to the project team, SCANS are "the workplace know-how that defines effective job performance today and therefore lies at the heart of job performance."

To order the HMMT National Voluntary Skills Standard, contact CORD at P.O. Box 21689, Waco, TX 76702-1689, or visit the CORD Web site at http://www.cord.org/.

Figure 1-1: SCANS Are Critical for the Success of Environmental Technicians

| Skills Employers Seek in College Graduates | |
|---|---|
| **Skill** | **Rating** |
| Oral communication skills | 4.7 |
| Interpersonal skills | 4.6 |
| Teamwork skills | 4.5 |
| Analytical skills | 4.4 |
| Flexibility | 4.3 |
| Leadership skills | 4.2 |
| Written communication skills | 4.2 |
| Proficiency in field of study | 4.2 |
| Computer skills | 4.1 |
| Five (5) is the highest value on a 1–5 scale. | |

Figure 1-2: *Community College Week* Survey (2/24/97) of Skills Employers Seek in College Graduates

# Checking Your Understanding

## Activity A

1. Answer the SCANS questions to analyze your foundation skills and workplace competencies. From your answers, identify your strong points as well as areas that need improvement.

Areas Needing Skills Development:

______________________________

______________________________

______________________________

______________________________

______________________________

2. Having identified skills needing improvement, establish goals for acquiring those skills. Develop and write a personal plan for achieving your goals.

## SCANS

**Directions:** *Each skill has been presented as a question to help you assess your areas of strength and opportunities for self-improvement. On a separate sheet of paper, answer each of the following questions.*

### Foundation Skills

#### Basic Skills

1. Reading – Can you understand and interpret written information?
2. Writing – Can you clearly communicate ideas and information in writing?
3. Arithmetic – Can you add, subtract, multiply, divide, compute decimals and fractions, and measure?
4. Mathematics – Can you select and apply appropriate math concepts to solve problems?
5. Listening – Do you pay attention and respond to verbal messages from others?
6. Speaking – Can you organize your ideas and communicate orally?

#### Thinking Skills

7. Creative thinking – Are you an idea person?
8. Decision-making – Do you consider information and issues when making choices or preparing to take action?
9. Problem solving – Can you determine solutions to problems?
10. Mental visualization – Can you see in your head how something might look or how a situation might unfold?
11. Knowing how to learn – Can you determine when information needs to be memorized or concepts need to be applied to real-world problems?
12. Reasoning – Can you arrive at a conclusion from a set of information?

#### Personal Qualities

13. Responsibility – Do you stick with a job until it is finished and try your best?
14. Do you accept a fair part of the blame when something does not go right?
15. Self-esteem – Do you have a healthy view of yourself?
16. Sociability – Do you treat others with kindness and caring?
17. Self-management – Do you maintain self control under pressure? Do you set realistic goals and meet them consistently?
18. Integrity/honesty – Do you choose ethical courses of action?

### Workplace Competencies

#### Managing Resources

19. Time – Do you use your time wisely, give your job a full day's effort, and take care of priorities?
20. Money – Do you operate within the budget, keep proper records, and help your company find ways to save money?
21. Material and facility resources – Do you properly store and maintain equipment and materials?
22. Human resources – Do you assess the skills of people you manage and distribute work with those skills in mind?

#### Using Information

23. Acquisition and evaluation – Do you use research sources such as the library, the Internet, professional journals, and personal contacts?
24. Organization – Do you keep good records? Can you find what you are looking for when you need it?
25. Interpretation and communication – Can you analyze data and explain its meaning to others?
26. Computer use – Can you use basic office programs such as word processing, database, and spreadsheet?

#### Interpersonal Skills

27. Teamwork – Can you contribute to a group effort and share leadership appropriately with all members?

28. Teaching - Can you help others learn needed knowledge and skills?
29. Service - Do you try to satisfy your client or customer?
30. Leadership - Do you motivate your colleagues? Do you offer constructive criticism and guidance? Do you encourage others to keep lines of communication open?
31. Negotiation - Are you flexible during decision-making situations?
32. Cultural sensitivity - Do you work well with people from diverse backgrounds?

### Systems

33. Comprehension - Can you identify the structure of an organization or technical system and describe how the parts of the structure interconnect and affect each other?
34. Performance monitoring - Can you determine if a part of a system is not performing, predict the results of the performance problem, and make the appropriate adjustments?
35. Improvement and design - Do you look for more efficient or cost-effective ways to do something and then make suggestions to change the system that is in place?

### Technology

36. Selection - Can you select appropriate equipment and methods for a task?
37. Application - For a selected technology, can you carry out the procedures and achieve the goal of the task?
38. Maintenance and trouble shooting - For a selected technology, can you do preventive maintenance, identify a problem, and determine solutions?

Achievement Plan for Improving Skills:

________________________________________

________________________________________

________________________________________

________________________________________

________________________________________

3. Be prepared to discuss in class how each SCANS skill is dependent on the ability to communicate.

## Activity B

This chapter discussed the importance of communication and its relationship to employment. Conduct your own research and determine what the business world considers the most important communication skill in the workplace. Suggested research media include:

Textbooks
Newspapers
Magazines
Internet
Personal interviews

Write a short paragraph summarizing your findings. Each class member will read their paragraph to the rest of the class, followed by a discussion of the similarities and differences of the classes' research. The class discussion should be focused on reaching a consensus, supported by research, as to the most important communication skill in the business world. This determination may be aided by listing and then ranking the class findings by importance.

# 1-2 What Is Effective Communication?

## Concepts

- Effective communication is communication that achieves its intended purpose.
- Communication can be personal or non-personal.
- Communication can be improved by analyzing its contents and delivery.

Since communication skills are critical to important employability skills such as teamwork, leadership, and critical thinking, let's begin by defining communication and effective communication as follows:

—**Communication** is the exchange of thoughts, opinions, or information using speech, signals, or writing.

—**Effective communication** is communication that achieves its intended purpose.

Communication is a complex process that uses one or any combination of speech, signals, and writing. For example, speaking to another person involves both verbal (word-based) and nonverbal (gesture- and expression-based) messages. Nonverbal communication, sometimes called body language, is unconscious communication resulting from the use of one or any combination of postures, gestures, and facial expressions.

Speech, signals, and written communication are either non-personal or personal. Examples of non-personal communication are messages from the mass media (such as newspapers, television, and radio) and the professional media (such as industry journals, conventions, and government reports). Non-personal forms of communication are frequently generated in the workplace in the form of reports, notices, instructions, and procedures. A way to determine whether non-personal communication is effective or not is to ask the following:

—Does the communication accomplish its intended purpose?

—Are the gestures and language appropriate for the intended audience?

—Are the originators of the communication credible in their field?

—Does the format do a good job of delivering the intended message?

Personal communication refers to exchanges among individuals, either one-on-one or in small groups. During the course of a workday, we may find personality is central to our ability to communicate when socializing, such as at lunch or on breaks. For example, people tend to listen to us more if we are friendly and even-tempered than if we are unfriendly and moody.

When personal communication is work-related, such as training a crew, analyzing a problem, giving instructions, or proposing a strategy, personalities continue to play a role in effective communication. To determine if personal communication is effective, analyze an interchange by asking the following questions:

—Are words used that make you uncomfortable?

—Are thoughts presented clearly?

—During verbal interchanges, does anyone ruin the conversation by dominating it with their own ideas?

—Do people appear to be listening to each other?

—In the nonverbal portion of a verbal interchange, does anyone use an expression that could have a negative impact on others?

—Do some listeners, through gestures or body language, make it obvious that they have little regard for the speaker and for what he is saying?

As you observe and begin to draw conclusions about effective communication, start analyzing your own communication skills. You might consider asking a friend to tell you what areas need improve-

| Verbal Actions | Nonverbal Actions |
|---|---|
| Writing and reading | Pictures and observation |
| Speaking and listening | Gestures and observation |

Figure 1-3: Verbal and Nonverbal Actions

ment. If you choose to do this, be open to receiving criticism and honest, but kind, when offering advice. Changing verbal and nonverbal habits can be a difficult but a worthwhile step toward improving your career.

## Checking Your Understanding

### Activity A

Read the following paragraph before answering the questions below:

A carcinogen is a substance capable of causing cancer in mammals. Chemicals, radiation, and certain viruses are cancer causing agents. Most cancers are induced by either synthetic or naturally occurring chemicals, which may include inorganic and organic compounds and hormones. A human carcinogen is further defined as a substance that induces cancer in humans.

1. Do you consider the above example to be personal or non-personal communication?
2. What is the intended purpose of this communication?
3. If you were the intended audience, would you consider the language level to be appropriate?
4. Is this an example of effective communication?

### Activity B

Read the following paragraph before answering the questions below:

Many alkylating agents (chemicals that can add alkyl groups to DNA) are carcinogens and/or mutagens. They form reactive carbonium ions, which react with electron-rich bases in DNA, such as adenine, guanine, cytosine, and thymine. This leads to either incorrect pairing of bases or chromosome breaks, causing growth and replication of neoplastic cells.

1. Do you consider the above example to be personal or non-personal communication?
2. What is the intended purpose of this communication?
3. If you were the intended audience, would you consider the language level to be appropriate?
4. Is this an example of effective communication?

# 1-3 Elements of Communication

## Concepts

- Some of the elements of communication are purpose (why a message is communicated), audience (who is listening to the message), vehicles (the means used to deliver the message) and barriers (outside forces that keep an audience from listening).
- Effective communication involves analyzing and making decisions about the previously-mentioned elements in the following ways:
  - **Purpose** Know what the real purpose of the message is (e.g., to persuade or inform).
  - **Audience** Know your audience and adapt your message to their particular needs.
  - **Vehicle** Choose the best medium (e.g., a report, personal conversation, or presentation) for conveying your message.
  - **Barriers** Troubleshoot and prevent possible communication barriers.

It is possible to improve the effectiveness of both written and oral communications by analyzing it for these components:

—Purpose and audience

—Communication vehicles

—Communication barriers

## Purpose and Audience

All communication has both a purpose and an audience. The purpose may be to describe a situation, explain a procedure, or persuade someone. An audience is one or more readers or listeners. It is usually easy in a work situation to recognize the purpose of and audience for a spoken or written message. Some examples:

—Describing a situation (purpose) to your supervisor (audience)

—Explaining a procedure (purpose) to a client (audience)

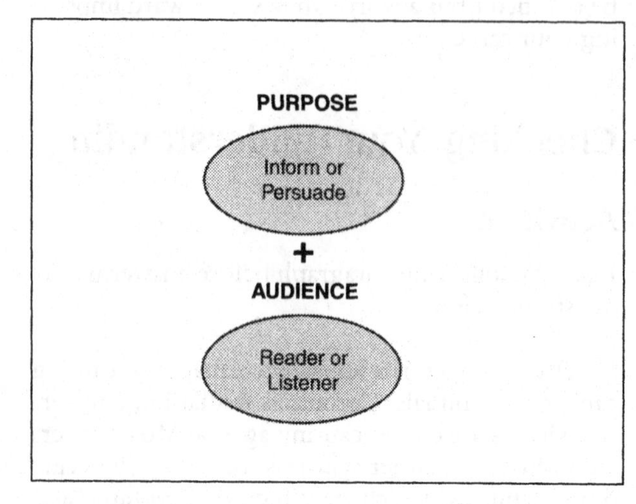


Figure 1-4: Effective Communication Requires a Purpose and an Audience

—Providing safety warnings (purpose) for people who use toxic substances in their jobs (audience)

Depending on audience characteristics such as age, educational level, experience, and attitude, you may need to adjust the content and style of the message. For example, the delivery of a message about a workplace rule regarding the necessity of washing hands may have to be different for lab technicians (who are exposed to toxic materials daily) and marketing executives (who are not).

## Communication Vehicles

A communication vehicle is the medium through which something is transmitted, expressed, or accomplished. There are many of them. For example, in the workplace a report may be the vehicle for sharing information, but its format may vary, consisting of one or more of the following:

—Verbal communication

—Written communication

—Visual communication

As an environmental technician, you could be called on to develop some or all of the communica-

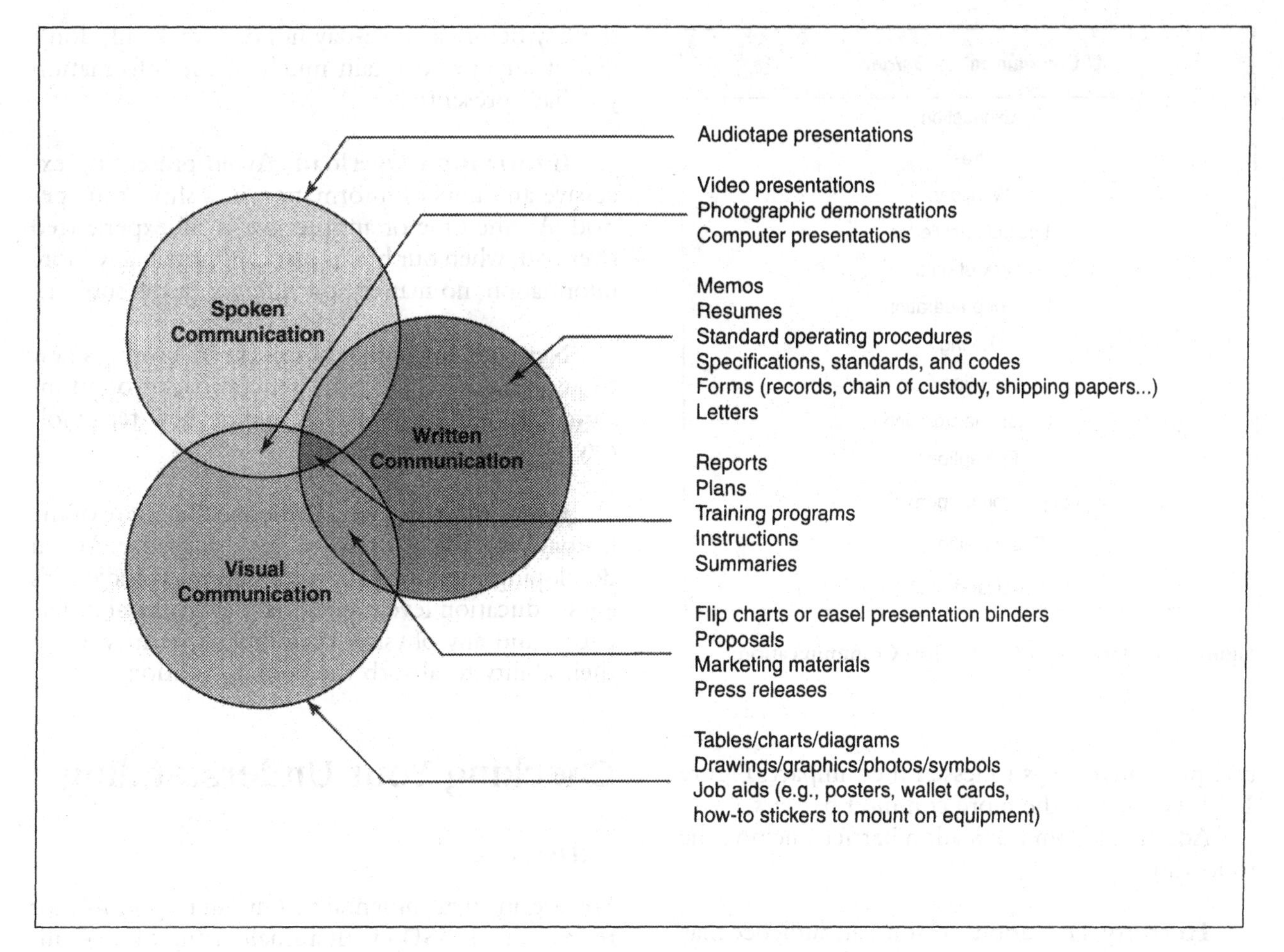


Figure 1-5: Communication Vehicles

tion vehicles listed in Figure 1-5. Although this list is not complete or in any particular order, it is a good representation of the diverse ways technicians may need to transmit information. Figure 1-5 also shows how the formats for each vehicle can vary and intersect. The methods used for developing several of these vehicles are introduced in Chapters 4, 5, and 6 of this handbook.

## Communication Barriers

Sometimes a message does not achieve its intended purpose because a barrier interferes or blocks it. Typical barriers include the use of unfamiliar terminology and failure to pay attention. Common barriers in the workplace are the presence of distractions and interruptions.

Since barriers prevent the achievement of the communication's purpose, try to anticipate them and

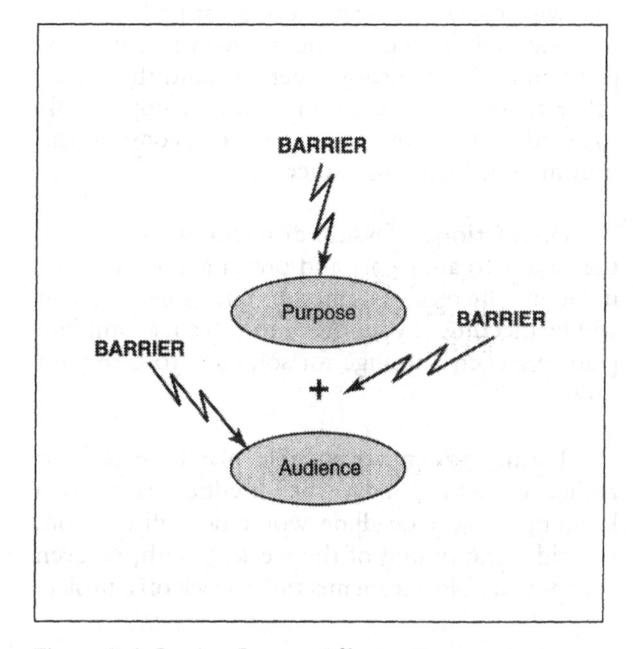


Figure 1-6: Barriers Prevent Effective Communication

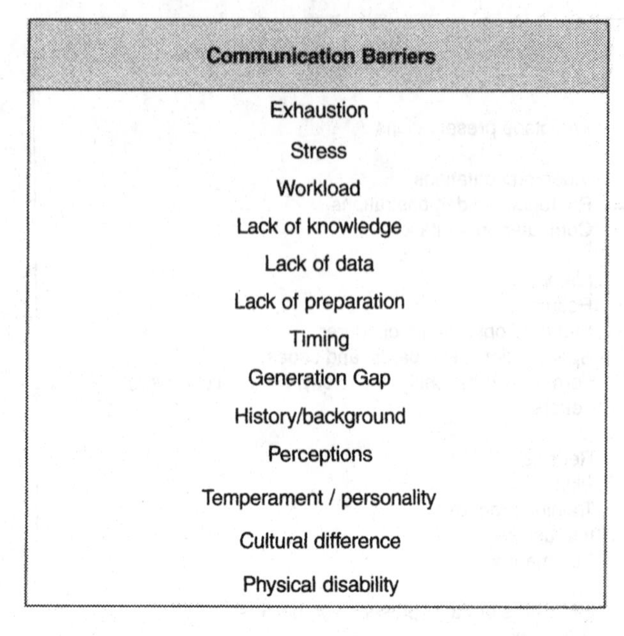

| Communication Barriers |
| --- |
| Exhaustion |
| Stress |
| Workload |
| Lack of knowledge |
| Lack of data |
| Lack of preparation |
| Timing |
| Generation Gap |
| History/background |
| Perceptions |
| Temperament / personality |
| Cultural difference |
| Physical disability |

Figure 1-7: Barriers to On-the-Job Communication

take preventive steps to lessen their impact. Figure 1-7 lists some of the more common barriers.

Additional communication barriers include the following:

**Hostility** For various reasons, an audience may be hostile toward either you or your message. For example, you may be the bearer of regulatory news that will cost the company money or perhaps have to communicate with someone who resents your position in the company. Keep in mind that inter-office tensions are common. You may not be able to avoid hostility, but you can select a communication method that will reduce it.

**Distractions** Physical distractions are perhaps the easiest to anticipate and prevent. For example, if you usually receive numerous telephone calls and will be meeting in your office to explain a sampling plan to a client, arrange for someone to take your calls.

**Timing** Whenever possible, plan to reach your audience at a time of day when tiredness, stress, or a looming project deadline won't be a distraction. Consider the timing of the week, month, or even year. If you hold a team meeting to kick off a project the day before a three-day holiday weekend, don't expect anyone to retain much of the information you have presented.

**Information Overload** Avoid presenting excessive amounts of information in a short time period. At one time or another we've all experienced overload, when our brains just can't retain any more information, no matter how interesting the subject.

**New or Unfamiliar Subjects** If a topic is new to an audience, start with general background information and, as it occurs, define new terminology using layman's terms.

**Unfamiliarity with Audience** Be aware of the history and background of your audience. When developing communication consider your audience's ages, education levels, personalities, cultural differences, and any physical disabilities that may affect their ability to absorb the communication.

## Checking Your Understanding

### Activity A

You are an environmental technician responsible for reviewing an MSDS with a small group of new employees. Write a brief description of the following:

—How you would decide which communication vehicle(s) to use for the presentation.

—What communication barriers you would anticipate.

—How, at the end of the presentation, you ensure that effective communication has taken place.

### Activity B

On a separate sheet of paper, develop a list of five different audience types, potential barriers to communicating with them, and strategies for overcoming those barriers. Start with the model provided (see next page).

Next, share your list with classmates, discuss each other's strategies, and create a new strategy list together.

| COMMUNICATION BARRIERS AND STRATEGIES FOR OVERCOMING THEM | | |
|---|---|---|
| AUDIENCE | BARRIERS | STRATEGIES |
| supervisor | distractions | close door/have calls answered |
| clients | unfamiliar subject | explain new ideas |
| | | |
| | | |
| | | |
| | | |
| | | |
| | | |

## Activity C

Prior to completing this activity, read Scenario 1 in Appendix 1.

As the new environmental technician for Locust®, you realize that you have a big job on your hands. The audience for your future communications will include the managers: Mr. Crabgrass, Mr. Green, Mr. Turf, and the first line workers.

What potential communication barriers, if any, do you anticipate with each of the managers?

# 1-4 Selecting an Appropriate Tone

## Concepts

- Tone is the selection of words and the manner in which they are expressed.
- Each business situation should determine a tone of formality or informality.
- All business communication should use a neutral tone that employs facts, not emotion.
- Business jargon should be used only when the audience is familiar with it.

Tone is the words you choose and manner in which you expression them. It can facilitate or block messages because it affects how the audience receives information. When speaking, we add tone to our communication through the words we choose and the following:

—Volume of our voice

—Vocal emphasis

—Facial expressions

—Gestures

In writing, tone becomes what the reader concludes by "reading between the lines." For example, it may be an attitude (e.g., friendly or sarcastic) that the reader senses.

**Formality**, **attitude**, and the **use of jargon** are all influenced by our personality, mood, and perception of the audience. All of these factors, therefore, tend to contribute to the tone of our conversation.

| Formal Tone | Informal Tone |
|---|---|
| Uses impersonal wording concerning people:<br>*""Each technician is responsible for submitting weekly status reports by 4:00 Friday afternoon.""* | Uses personal wording such as "you" or names:<br>*"Each of you is responsible for submitting weekly status reports by 4:00 Friday afternoon."* |
| Avoids contractions:<br>*"The status reports should not be longer than one page."* | Accepts contractions:<br>*"The status reports shouldn't be longer than one page." or "Let's keep the reports to a page."* |
| Uses multiple clause sentences (that, is several thoughts joined by connecting words):<br>*"After the technician writes the status report, he or she shall complete the weekly time sheet, which provides a means of documenting chargeable periods spent on each client's site as well as the nonchargeable periods, such as the minutes spent traveling from site to site - other words, the amount of time considered to be allotted to the firm's overhead costs."* | Uses conversational sentence structure:<br>*"After you write the status report, complete the weekly time sheet. The time sheet will help us keep track of time spent on each client's site. It will also help us get a feel for how much time isn't able to be billed to our clients. These figures will be a useful part of figuring out our overhead."* |
| Adheres to traditional rules of grammar:<br>*"The supervisor to whom each technician submits his or her status report and time sheet shall check off the employee's name upon receipt of the materials."* | Relaxes some traditional rules of grammar:<br>*"The supervisor you report to will check your name off as you hand the materials over."* |
| Selects complex words:<br>*"The supervisor shall furthermore maintain all materials in a chronologically arranged file."* | Selects simple words:<br>*"The supervisors will file the materials by date"* |

Figure 1-8: Formal versus Informal Tone

## Formality versus Informality

Deciding whether communication tone should be formal or informal often depends on the audience and purpose. Two examples of situations requiring greater formality are a carefully worded response to a government inspector's report and a business letter to someone you do not know. An informal tone may be used, for example, in a memo to a fellow project team member to schedule a team meeting. Figure 1-8 gives some examples of formal and informal tone.

As shown in Figure 1-9, formality of tone can be viewed as a continuum with impersonal wording and complex jargon on the extreme left and emotional wording and slang on the extreme right. Most business communications should be targeted somewhere in the middle.

## Attitude

Communication tone is also conveyed by the speaker's attitude. Tone of voice (e.g., agitated, anxious, happy) can reveal attitude as much as words. Voice inflection also reflects attitude. For example, using only vocal inflections, see how many different attitudes you can communicate using the following words:

"You're the expert"

Since writers cannot rely on visual cues to convey attitudes, tone is dependent upon the sequence of words chosen. Most business writing uses a "neutral tone" where the wording does not reveal the writer's mood or personality.

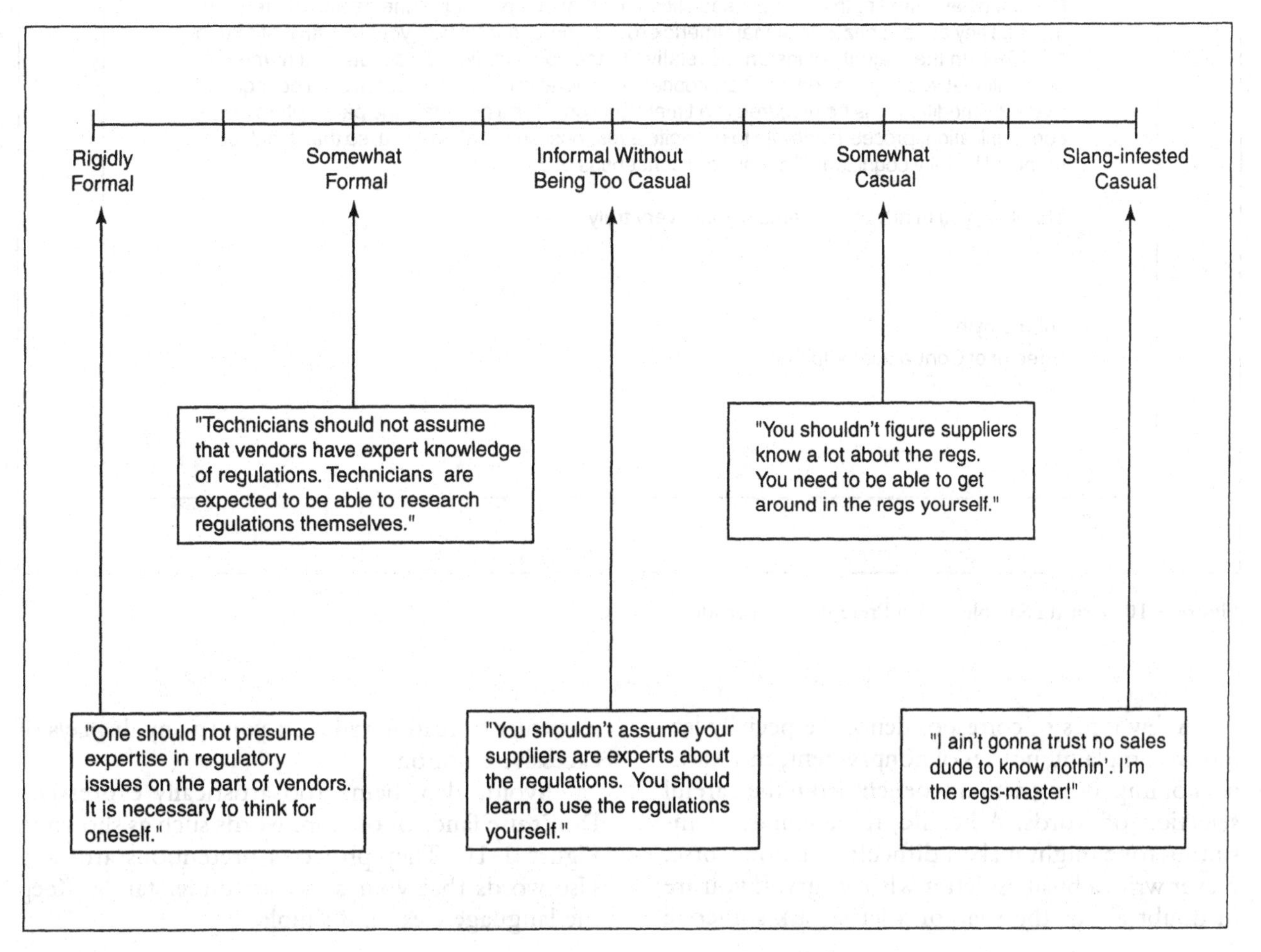


Figure 1-9: Formality Continuum

*Smugly, Smugly, and Pompus*
*5000 Superior Boulevard*
*Garden City, Virginia 22432*

March 19, 1999

Mrs. Jane Johnson
100 Hollyhock Lane
Garden City, Virginia 22432

Dear Madam:

Enclosed herewith are the contract specifications that comprise all of the necessary requirements. They are prioritized in alphanumeric order commensurate with your directive of March 15, 1999. In the majority of instances relative to the specifications, a substantial number of other alternatives might be deemed appropriate for utilization; in which case, it shall be required that such modifications be procured in a timely fashion. If you ascertain the parameters of this communication, proceed forthwith to expedite a response in the affirmative, so that this department will be fully cognizant of our shared consensus of opinion.

Thanking you in advance, I remain yours very truly,

John Smyth
Director of Contractual Affairs

Figure 1-10: Writing Sample with a Pretentious Attitude

A few types of correspondence – especially letters of complaint, notices of nonpayment, and warnings of impending action – benefit from the careful selection of words. A hostile, threatening, or insulting tone might make a difficult situation worse. Never write a business letter when angry. If you are in doubt about the tone of a letter, ask someone you trust to read it and alert you to any displays of excessive emotion.

Avoid, also, being too artistically expressive. Don't use fancy or obscure words such as shown in Figure 1-10. They project a pretentious attitude. Use words that your audience understands. Keep the language clear and simple.

# Jargon

Whether you call it jargon or job talk, the specialized terminology of your work may be confusing to those unfamiliar with the field. Figure 1-11 contains the same information worded first to accommodate a professional audience and then to accomodate a lay audience. Use of special terminology may strike your reader or listener as overly formal and difficult. For instance, to the person outside the wastewater treatment field the industry jargon text in Figure 1-11 has no meaning and may appear as too formal. To the wastewater treatment professional, however, the passage is understandable and moderate in tone.

Some tips regarding the use of jargon include the following:

—To determine how much jargon to use, analyze the level and needs of the audience.

—Remember that environmental concepts have their origins in the regulatory arena, which can be new and difficult to communicate to clients, support staff, and other non-technical people.

—Use terminology that is appropriate for the audience. This includes acronyms, which are words formed from initials, such as RCRA for Resource Conservation and Recovery Act.

—If new terminology must be incorporated, then define it.

—If speaking, watch for gestures indicating confusion. Your awareness of the needs of others will help you know when to clarify information.

—If you are communicating through writing, then ask someone at the level of your intended audience to review your message and to identify areas that may need further clarification or explanation.

# Selecting Tone

Writers and speakers should base their selection of tone on the needs of the audience, the purpose of the communication, and the situation. When selecting the proper tone, ask the following questions:

—Should the tone be formal or informal?

—Is humor appropriate?

—What level of terminology or jargon is appropriate for the audience?

| Upflow Anaerobic-Sludge Blanket (UASB) Reactors Description | |
|---|---|
| Industry Jargon | Layman's Terms |
| In the UASB reactor, the input waste stream is from the bottom and it passes through a blanket of sludge granules as it ascends. The granules consist of the active anaerobic biomass and tend to entrain biogas and become buoyant. An overhead baffle system is located at the top of the reactor to capture and promote degassing of the granules. The granules subsequently settle to the sludge blanket, and in this way the biomass of the reactor is efficiently controlled. Maintaining the organisms while treating high COD waste streams is common. | In an upflow anaerobic-sludge blanket (UASB) reactor, the wastewater enters at the bottom where it encounters a large number of suspended particles known as the anaerobic-sludge blanket. The blanket is composed of clusters of anaerobic bacteria that feed on the nutrients in the wastewater. The gas generated by the bacteria's feeding frenzy forms and traps tiny bubbles in the cluster that causes it to slowly float upward. At the top of the reactor, a series of baffles cause the bubbles to separate from the cluster, allowing it to settle and rejoin the sludge blanket until it experiences another cycle. This cyclic process maintains the number of bacteria in the reactor, even when high chemical oxygen demand (COD) waste streams are introduced. |

Figure 1-11: Upflow Anaerobic-Sludge Blanket (UASB) Reactors Description in Industry Jargon and in Layman's Terms

—Is an overview sufficient or is more detail necessary?

—What does the setting for your readers or listeners suggest – a serious, no-nonsense approach or a lighter, interest-engaging approach?

## Checking Your Understanding

### Activity A

1. Analyze the following paragraph and determine if it incorporates all the points made in Figure 1-10:

   Enclosed is a draft of the contract specifications we discussed on March 14, 1999. Please look them over and verify that they include your requirements and are organized as you have wished. Possible alternatives are available for many of the materials specified, but I would need to research them quickly to be able to include them before the contract is to be signed. If you approve of the draft as it is, please sign and return a copy of this memo for our records.

### Activity B

1. Rewrite the following sentence to make the language clearer:

   One is advised that verifying regulatory text is preferred over relying purely upon the ability to recall the content of the verbiage.

2. Reduce this sentence to a four-word sentence common in the environmental industry:

   Claiming a lack of exposure to requisite information regarding environmental legislative edicts fails to serve as a justification for errant corporate behaviors.

3. Replace the following sentence with a popular two-word phrase:

   Management stipulates that employees refrain from bewailing their situations.

4. Use vocal tone and inflections to express the following sentence as a) a positive statement, b) an expression of surprise, and c) a sarcastic comment.

   "You presented a professional image today."

5. Make a list of nonverbal cues that may indicate that a person is not listening to what you are saying.

### Activity C

With your class, develop a list of verbal and nonverbal communication styles that you all think are most appropriate in a business environment. For example, tone of voice, gestures, choice of words, etc.

### Activity D

Prior to completing this activity, read Scenario 1 in Appendix 1.

As the new environmental technician for Locust®, you realize that you have a big job on your hands. The audience for your future communications will include the managers: Mr. Crabgrass, Mr. Green, Lou Turf, and the first line workers.

What kind of jargon and tone would be most effective for communicating with the first line workers?

## Summary

The ability to communicate is central to job success. Communication is the exchange of thoughts, opinions, or information using speech, signals, or writing. Effective communication is communication that achieves its intended purpose.

There are numerous vehicles by which to communicate ideas and information. Among the vehicles that are widely used in environmental work are reports, plans, business letters, instructions, standard operating procedures, and memos.

Effective verbal and written communication requires the following:

—Determining the purpose for writing, speaking, or illustrating. Deciding whether you want to inform or persuade

—Adjusting the content and style of your message according to the characteristics and needs of your audience.

—Predicting and preventing communication barriers, such as distractions and information overload.

—Avoiding wordiness, slang, emotional sentiments, and the unnecessary use of big words.

—Selecting the appropriate tone for your purpose and audience - formal or informal, technical or non-technical.

PRESERVING THE LEGACY

2

# Writing and Technical Writing Basics

## Chapter Objectives

Upon completing this chapter, the student will be able to:

1. **Explain** the organizational structure used in the Code of Federal Regulations (CFR).
2. **Explain** the differences between specifications, standards, and codes.
3. **Evaluate** an information source for scientific validity.
4. **Evaluate** an Internet source.
5. **Understand** how to develop an outline.
6. **Employ** the use of headings, lists, and illustrations in a document.
7. **Edit and proofread** a document by correcting its sentence structure, grammar, and punctuation.

## Chapter Sections

# 2-1 Writing Procedure

## Concepts

- Depending on the job requirements, environmental technicians can be expected to write a wide variety of technical documents.
- The general procedure for writing technical documents includes brainstorming, researching, outlining, writing, editing, and proofreading.

In addition to traditional business correspondence, environmental technicians may be asked to write or complete the following:

—Requests to manufacturers for Material Safety Data Sheets (MSDS) (see Chapter 3)

—Reports (see Chapters 3 and 4)

—Instructions (see Chapters 3 and 4)

—Standard operating procedures (see Chapter 4)

—Hazardous communications plans (see Chapter 4)

—Log book data entries (see Chapter 4)

—Field notes (see Chapter 4)

—Environmental compliance form entries (Chapter 5)

Although requirements differ for every business and environmental document, the writing process is as follows:

1. Brainstorming ideas and issues related to the topic
2. Grouping the ideas and issues into subtopics
3. Researching and gathering information
4. Outlining the document's structure
5. Writing the document
6. Editing and proofreading

Brainstorming is the first step in the writing process. It involves listing ideas for a topic without stopping to think about them. By striving to generate as many ideas as possible, you are engaging in "fluent thinking." One idea can lead to another, especially when more than one person participates. During brainstorming, participants hold their judgment about the worth of ideas being generated. This encourages creativity, a state referred to as "flexible thinking." Creative thinking makes most of us feel good, resulting in a positive attitude toward writing a document – the best preventive medicine for writer's block.

The next step in the writing process is organizing all of the brainstorming ideas and issues into groups of subtopics related to the main topic. If the subtopics need to be substantiated or require additional information, then you must move on to the third step: researching and gathering information. Section 2-3 through section 2-6 describe the main research tools used by environmental technicians for writing technical documents. Section 2-3 shows you how to use the Code of Federal Regulations (CFR) to ensure that your standards and procedures comply with the law. Section 2-4 introduces specifications, standards, and codes. In section 2-4 you will learn how to determine whether the information you find in your research is scientifically valid. Finally, section 2-5 provides guidelines for using the Internet as a research tool for scientific information.

Hint: To make outlining easier, write your research notes on index cards.

The last three sections of this chapter describe how to use an outline for organizing a document, how to write a document, and how to edit and proofread.

# 2-2 Research: Using the Code of Federal Regulations (CFR)

## Concepts

- Environmental technicians must refer to the Code of Federal Regulations (CRF) to ensure that written standards and procedures comply with federal law.
- The CFR is a hierarchy organized, from top to bottom, by titles, chapters, parts, sections, and paragraphs.

The CFR is an important reference for writing standard operating procedures, technical instructions, and hazardous communication plans. Environmental technicians use it to ensure that their procedures and plans follow the law. The CRF is a set of books printed annually by the federal government that contain all federal regulations. It is also available from private vendors such as Solutions Software Corporation (http://www.env-sol.com) on CD-ROM. A number of other vendors can be found on the Internet by using one of the search engines and entering CFR+CD+ROM. The National Archives and Records Administration maintains a searchable database at http://www.access.gpo.gov/nara/about-cfr.html and it is free. In addition to the CFR is a daily publication called the *Federal Register*, which provides the latest information regarding proposed and new regulations as well as public notices concerning regulations.

Everyday, the *Federal Register* announces the notices, background information, and legislation of our government. Environmental personnel use the Federal Register to keep up on the latest regulations.

Annually, the *Code of Federal Regulations* (CFR) publishes all of the rules that carry out the intent of U.S. legislation. Environmental personnel use the CFR to develop policies and procedures that ensure their agencies or companies comply with the law.

To find specific information within this extremely large document set, it is helpful if you understand how it is structured. Its levels, from highest to lowest, are as follows:

—**Title** A title covers one broad topic and is designated by Arabic numbers. There are 50 titles representing the regulated areas. Examples of title numbers and names are 49 CFR (transportation), 29 CFR (labor), and 40 CFR (protection of environment).

—**Chapter** A chapter divides titles by federal agencies that administer the regulations. It is designated by Roman numerals. Note that chapters are not referenced in citation names (e.g., 29 CFR 1910.120). They are used only when speaking in general. For example, Title 29, Chapter XVII refers to the Occupational Safety and Health Administration regulations under Title 29.

—**Part** A part is a body of regulations within a title. It is designated by Arabic numbers. For example, 29 CFR 1910 indicates part 1910 of title 29 that deals with occupational safety and health standards. Subparts may also be used. They are designated by capital letters. Note that subparts are not referenced in citation names. They are used only when speaking in general. For example, 29 CFR 1910, Subpart H refers to hazardous materials.

—**Section** A section is a further division of a part. It is designated by an Arabic number that appears after a decimal point to the right of the part number. For example, 29 CFR 1910.120 indicates section 120 of part 1910 of title 29. The section is the basic unit of CFR organization.

—**Paragraph** A paragraph is a division of a section. It is designated as a lower case letter enclosed by parentheses. For example, 29 CFR 1910.120 (a) indicates paragraph a of section 120 of part 1910 of title 29. Paragraphs may be further divided using the following outline hierarchy:

| | |
|---|---|
| (1), (2), (3), etc | 29 CFR 1910.120 (a)(1) |
| (i), (ii), (iii), etc. | 29 CFR 1910.120 (a)(1)(ii) |
| (A), (B), (C), etc.* | 29 CFR 1910.120 (a)(1)(ii)(A) |
| (1), (2), (3), etc. | 29 CFR 1910.120 (a)(1)(ii)(A)(3) |
| (i), (ii), (iii), etc. | 29 CFR 1910.120 (a)(1)(ii)(A)(3)(iii) |

* Note that sometimes lower case italic letters are used for this subdivision

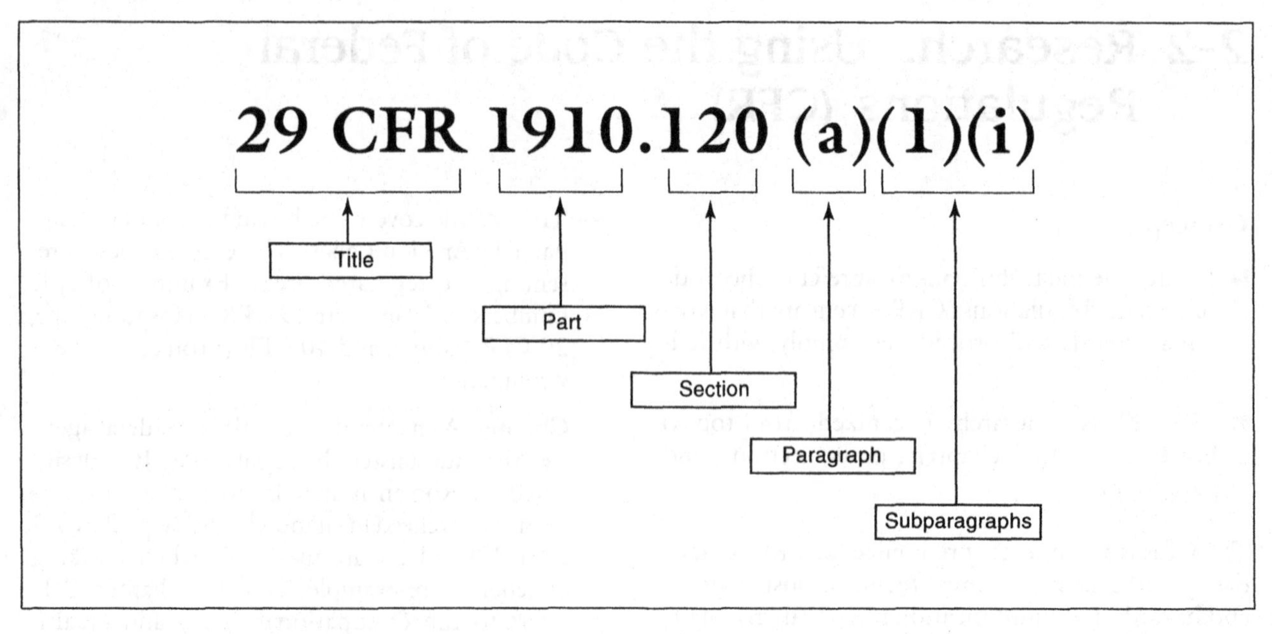


Figure 2-1: Dissection of a CFR Citation

| 29 CFR 1910.120(a)(1)(i) | | |
|---|---|---|
| **Citation Component** | **Name of Component** | **Actual wording of the component in the example citation** |
| The first three components below represent the basic numeric coding. | | |
| **29** | Title | Labor |
| **1910** | Part | Occupational Safety and Health Administration, Department of Labor |
| **120** | Section | Hazardous Waste Operations and Emergency Response |
| The following components represent the outlining levels within a regulation. | | |
| **(a)** | Paragraph | Scope, application, and definitions -- |
| **(1)** | First level subdivision of a paragraph | Scope. This section covers the following operations, unless the employer can demonstrate that the operation does not involve employee exposure or the reasonable possibility for employee exposure to safety or health hazards: |
| **(i)** | Second level subdivision | Clean-up operations required by a governmental body, whether federal, state, local or other involving hazardous substances that are conducted at uncontrolled hazardous waste sites (including, but not limited to, the EPA's National Priority Sites List (NPL), state priority site lists, sites recommended for the EPA NPL, and initial investigations of government identified sites which are conducted before the presence or absence of hazardous substances has been ascertained); |
| **(A)** | Third level subdivision | The example above does not contain a third level subdivision. |
| **(2)** | Fourth level subdivision | The example above does not contain a fourth level subdivision. |
| **(i)** | Fifth level subdivision | The example above does not contain a fifth level subdivision. |

Table 2-1: Components of a Citation for an OSHA Regulation

Table 2-1 shows the text from the CFR describing 29 CFR 1910.120 (a)(1)(i).

Finding detailed citations in the CFR can be a challenge because the indentations associated with a traditional outline are omitted on the CFR pages and outline letters and numbers do not show up very well on a page. Some readers devise their own systems for deciphering the citations, such as marking the alphanumeric numbering with highlight markers or symbols like circles and squares. If using the Internet or a CD-ROM, take advantage of your particular software's "search" or "find" capabilities to look up key words.

Should interpretation of the regulations become difficult, seek assistance from the government agencies themselves, legal professionals, or professional organizations whose purpose is to enlighten their members through journals and newsletters.

## Checking Your Understanding

### Activity A

Photocopy the portion of OSHA's Respiratory Protection standard (29 CFR 1910.134), which is included in Appendix 3. Devise a system to code the main topics and supporting details by using different colored pencils or highlighting markers.

### Activity B

After performing Activity A, write a paragraph describing the regulation's main ideas to an audience who knows nothing about respiratory protection.

# 2-3 Research: Understanding Specifications, Standards, and Codes

## Concepts

- How specifications, standards, and codes differ.
- How the meanings of the words shall and should differ within standards and codes.
- Copyright rules regarding the reprint of standards.

In addition to the CFR, environmental technicians use specifications, standards, and codes as reference tools. Often these three entities are mistaken for one another.

**Specifications** are descriptive statements of quality or performance that a buyer can expect from a manufacturer's materials and/or equipment. Specifications for the same type of product may vary from company to company. For example, the specifications for cameras from two different manufacturers will not be the same.

**Who needs to know about specifications?** Individuals who purchase equipment. Specifications provide critical information about product quality and performance.

**Who needs to know about standards?** All of us. Standards affect many parts of our lives. Imagine the problem if the manufacturer of your computer's CD-ROM did not observe the agreed-upon industry standards. The player probably would not work in your computer. Imagine if the manufacturer of your extension ladder cut corners on the quality of materials and strength of construction.

**Who needs to know about codes?** People who work in an industry that in any way affects health, safety, and the environment. This includes those who are protected by adherence to the codes and those in jobs in which they are obligated by law to protect others. Ignorance of laws and codes is rarely excused by enforcement agencies. If you assume professional, supervisory, or ownership responsibilities, it is essential that you research all the regulations that apply to your situation.

**Standards** are specifications that a group of interested persons in a particular field come to agree should be the recommended practice throughout an industry. Standards generally result from long studies and testing before they are approved for publishing and dissemination. A standard that is developed by a well-respected organization, such as the National Fire Protection Association (NFPA), is likely to be used very widely and consistently. For example, the food industry often uses the NFPA standards for installing equipment that removes the smoke and grease-laden emissions of commercial cooking equipment.

**Codes** are standards that are adopted as municipal, state, or federal regulations. Codes have the effect of law and must be followed by industries and individuals in the jurisdiction that adopts them. One challenge an environmental technician faces is finding all of the codes that apply to a project. For example, to transport and store hazardous materials you need to know about codes regarding the transportation of hazardous materials, local fire codes that apply to the storage of flammable materials, and codes covering worker safety.

To find out about the existence of codes and the details regarding them, check with the appropriate divisions of city, county, state, and federal agencies. Professional organizations are good resources too. Also, try searching the Internet. For example, many cities have their building codes available on the Internet or have adopted the Uniform Building Code, which is published by an organization called the International Conference of Building Officials that maintains a web site at http://www.codes.org. A final resource is an attorney that specializes in regulatory law.

code (kod) n., 1. A systemically arranged and comprehensive collection of laws....

*The American Heritage Dictionary*

An exception to the above definition of "standards" is OSHA's use of that word in the CFR. Part 1910 of title 29, entitled "Occupational Safety and Health Standards," describes regulations, not recommended practices.

Figures 2-2 through 2-5 demonstrate the similarities and differences among specifications, standards, and codes. Notice the use of the words shall and should. Shall indicates a mandatory direction; should indicates non-mandatory guidance.

Figure 2-3 shows a brief excerpt from NFPA 560, which is the National Fire Protection Association's standard for Storage, Handling, and Use of Ethylene Oxide for Sterilization and Fumigation, 1995 Edition. This is NFPA's recommended practice.

**Typical Materials Specification for a Fume Hood Alarm**

All fume hoods shall come equipped with a Suck-It-Up fume hood velocity alarm. The units shall be flush-mounted in a fume hood side fascia panel. Any system that cannot be flush-mounted in this manner is unacceptable. The velocity alarm shall signal an unsafe operating condition when the fume hood face velocity falls below a pre-set amount. The alarm set-point calibration is performed by the user\owner once a proper face velocity has been set and measured. The alarm system shall consist of the following:

a. A 12-light LED display that registers face velocities between 0 and 165 FPM by an increasing number of illuminated LED's for larger face velocities.

b. An interconnected set of 12 micro-switches that serves as a way of selecting any of the 12 LED display lights as the alarm set-point.

c. A buzzer alarm of at least 85 decibels.

d. A flashing warning light in synchronization with the audible alarm.

e. A shut-off switch for the audible alarm that will not stop the warning light.

f. A test mode that simultaneously tests LED function and alarm set-point.

g. A velocity alarm system furnished complete with velocity detector, detector mounting hardware, alarm, and optional case. The system shall operate at 120V, 60hz power.

h. An alarm front panel and steel enclosure box made of 18-gauge stainless steel.

Figure 2-2: Example of a Specification

**Chapter 5 Gas Dispensing Areas**

**5-1 General.**

In addition to the requirements in Chapter 3, Storage of Ethylene Oxide, the following shall apply to areas where ethylene oxide is dispensed from containers. Ethylene oxide storage shall be permitted in dispensing areas.

**5-1.1**

Indoor dispensing areas shall be equipped with a continuous gas detection system that provides an alarm when ethylene oxide levels exceed 25 percent of the lower limit of flammability (7,500 ppm).

NOTE: Additional detection at lower levels may be required to meet the requirements of the Occupational Safety and Health Administration of the U. S. Department of Labor (29 CFR 1910.1047).

**5-1.2**

Exhaust ventilation shall be installed in all indoor dispensing areas used for ethylene oxide. Exhaust ventilation shall comply with the following:

(a) Mechanical ventilation shall be operated continuously at a rate of not less than 1 $ft^3$ per min per $ft^2$ (0.3 $m^3$ per min per $m^2$) of floor area of dispensing area.

(b) Exhaust ventilation shall not be re-circulated within a room or building.

*Exception: Where the air is treated to reduce the ethylene oxide concentration to below that which represents a hazard, re-circulation shall be permitted. Controls shall be provided to ensure the performance of the treatment and re-circulation system.*

(c) The ventilation system shall be designed to prevent accumulation of ethylene oxide anywhere in the dispensing area.

(d) Loss of ventilation shall activate a visual and audible alarm and shall stop the flow of ethylene oxide at the remotely operated shutoff valve closest to the container.

[And from the non-mandatory Appendix A that follows NFPA 560...]

**A-5-1.2**

Local exhaust hoods are an effective means used to control ethylene oxide levels at the source of potential release.

Reprinted with permission from NFPA 560 *Storage, Handling, and Use of Ethylene Oxide for Sterilization and Fumigation*, Copyright © 1995, National Fire Protection Association, Quincy, MA 02269. This reprinted material is not the complete and official position of the National Fire Protection Association, on the referenced subject which is represented only by the standard in its entirety.

Figure 2-3: Excerpt from the NFPA Standard

The Chemical Hygiene Plan...shall include each of the following elements:

...the selection of control measures for chemicals that are known to be extremely hazardous....

...a requirement that fume hoods and other protective equipment are functioning properly....

...provisions for additional employee protection for work with particularly hazardous substances. These include 'select carcinogens,' reproductive toxins and substances which have a high degree of acute toxicity. Specific consideration shall be given to the following provisions which shall be included where appropriate:

a. Establishment of a designated area;
b. Use of containment devices such as fume hoods or glove boxes;
c. Procedures for safe removal of contaminated waste; and
d. Decontamination procedures.

Excerpt from an appendix within an OSHA Standard

[1910.1450 includes Appendix A, which the CFR refers to as nonmandatory recommendations. Nonmandatory means this is NOT a regulation. The Appendix includes this recommendation for fume hoods:]

A laboratory hood with 2.5 linear feet of hood space per person should be provided for every 2 workers if they spend most of their time working with chemicals; each hood should have a continuous monitoring device to allow convenient confirmation of adequate hood performance before use. If this is not possible, work with substances of unknown toxicity should be avoided or other types of local ventilation devices should be provided.

Figure 2-4: Excerpts from an OSHA Standard

**505.7 Hoods and Enclosures.** Hoods and enclosures shall be used when contaminants originate in a concentrated area. The design of the hood or enclosure shall be such that air currents created by the exhaust systems will capture the contaminants and transport them directly to the exhaust duct. The volume of air shall be sufficient to dilute explosive or flammable vapors, fumes or dusts as set forth in Section 505.4. Hoods of steel shall have a base metal thickness not less than 0.027 inch (0.68 mm) (22 gage) for Class 1 and Class 5 metal duct systems ...[List of metal thicknesses for Classes 2 - 5]...Approved nonmetallic hoods and duct systems may be used for Class 5 corrosive systems when the corrosive mixture is nonflammable. Metal hoods used with Class 5 duct systems shall be protected with suitable corrosion-resistant material. Edges of hoods shall be rounded. The minimum clearance between hoods and combustible construction shall be the clearance required by the duct system.

*For brevity, the list of metal thickness for Classes 2 – 5 has been omitted.

Figure 2-5: Excerpt from the Uniform Mechanical Code

Figure 2-4, an excerpt from 29 CFR 1910.1450, OSHA's Chemical Hygiene Standard, represents a trend in government regulations toward the use of "performance standards." The performance standard sets goals for the employer to meet, such as those in the example here. The employers then decide individually how to meet those goals.

An example of how an environmental technician would use specifications, standards, and codes is a written report recommending the purchase of an engineering control technology that provides the best level of indoor air quality for the lowest purchase and maintenance cost. Standards and codes would be the resources used for finding data about the substances involved. For example, if the substances are ethylene oxide and lead, it would be useful to know the following:

1. Ethylene oxide is a flammable and highly toxic gas with an OSHA exposure limit of one part ethylene oxide per one million parts air (1 ppm).
2. The lead used in the operations will produce fumes and dust.

Specifications would be used to analyze and compare a range of commercial ventilation systems designed to remove chemical emissions and airborne particles.

## Copyright Regulations

Documents published by government agencies are considered to be public domain, which means that individuals may reproduce those materials without seeking permission. For example, you could legally reproduce material from the CFR for an employee handbook or articles from the EPA Journal for a company magazine. Note, however, that the sources for any public domain information must always be cited.

Since standards are developed at great expense by organizations, they tend to copyright their standards publications and often charge a fee for permission to reproduce material. These fees are used by professional organizations to continue their research. There are several ways in which standard data and information can be used. As an example, consider the following options for using the National Fire Protection Association (NFPA) hazardous materials emergency response standards as part of a company-wide training program:

1. Purchase the appropriate number of copies from the NFPA to distribute to each trainee.
2. Seek permission from NFPA to reprint the standards in your written training program, and pay NFPA's permission fee.
3. Purchase a few copies from NFPA that trainees may check out. Cite the title of the standards in your written training program.
4. Quote very briefly from a standard, as done in this chapter, and credit the source. The U.S. Copyright Act includes fair use clauses that permit very limited reproduction. For more information refer to this handbook's bibliography.

Buying one copy of a private organization's standards book and reproducing a set for trainees is an infringement of the copyright law and illegal. Note also that standards referenced in public domain documents are still protected by copyright law and require reprint permission.

## Checking Your Understanding

### Activity A

Interview someone whose work involves the use of specifications, standards, and/or codes. Possible candidates include an environmental consultant, architect, engineer, firefighter, building inspector, construction superintendent, police, or manufacturer. Prepare a brief written report about how specifications, standards, and/or codes are used in his or her profession and share your findings with the class.

The following are some suggested interview questions:

1. What are the specifications, standards, or codes that influence your work?
2. What groups contribute to the development of the standards or codes that you use?
3. Once standards or codes are developed and in use, are they likely to change?
4. Do you ever write material or performance specifications?
5. Does the CFR influence your work? If so, what parts?
6. What role, if any, does a mechanical drawing play with respect to specifications?

### Activity B

While working for Locust® (see Scenario 1 in Appendix 1) you observe that some of the Personal Protective Equipment (PPE), such as hearing protectors, dust masks, canister-type respirators, and safety glasses, are in bad repair and dirty. You will be buying new PPE supplies for the employees.

1. For any one type of PPE, check the OSHA requirements in 29 CFR Part 1910. Then write a memo to your boss, Mr. Crabgrass, summarizing the PPE requirements for one of the following:
   - 29 CFR 1910.139 on respiratory protection (check the CFR)
   - 29 CFR 18910.95 on hearing protection (check the CFR)
   - 29 CFR 1910.133 in eye and face protection (check the CFR)
2. For certain Locust® operations, you think employees need dust masks or canister-type respirators. Check these CFR standards and write a summary regarding any requirements:
   - 29 CFR 1910 Subpart Q on welding, cutting, and brazing
   - 29 CFR 1910.107 on spray finishing using flammable and combustible liquids
3. For the PPE item you selected, use a safety equipment catalog to determine which professional standards organization, if any, has approved the models. For example, the National Fire Protection Association (NFPA), the American National Standards Institute (ANSI), etc. Look for the specifications included for each item. Add your findings to the memo for Mr. Crabgrass.

Section 2-3 was adapted from materials in *Living and Writing Responsibly: Environmental Readings for Classic and Technical Composition.* Raleigh: HMTRL, 1992, by permission of the publisher.

# 2-4 Research: Evaluating Sources and Scientific Validity

## Concepts

- The validity of published research should be based on the credibility of the source, the research's scientific validity (i.e., the research must have been obtained using the scientific method), and the rigor with which the scientific method was applied.
- Data can be manipulated by massaging, extrapolating, smoothing, slanting, fudging, manufacturing.

Headline from a peer-reviewed journal:

"Insidious Effects of a Toxic Estuarine Dinoflagellate on Fish Survival and Human Health," *Journal of Toxicology and Environmental Health*

Headline from the mainstream press:

"The 'Cell From Hell' – Pfeisteria Strikes Again in the Chesapeake Bay," *Newsweek Magazine*

Headline from a tabloid:

Pfeisteria Hysteria is Traced to Alien Plot," *Unnamed Tabloid*

**Peer-reviewed journal:** A professional journal that only accepts articles reviewed by a panel of readers who hold equal standing to the author. For example, scientists review articles by other scientists for the purpose of publishing.

**Mainstream press:** The generally accepted news organs (i.e., magazines, newspapers, and television) that the public believes for the most part to contain true stories and information.

**Tabloid:** A newspaper-like publication that often uses sensational subjects and writing. Such publications are usually not linked with truthfulness.

Articles on the environment run the gamut from those that are scientifically documented to unsupported opinions. We immediately know to question the validity of a scientific article published in a supermarket tabloid, but not in a professional or peer-reviewed journal such as the *New England Journal of Medicine*. As for the mainstream media, how much can we trust the validity of science information reported in the local newspaper or on television's network news? We know that news services like the Associated Press (AP) strive to offer accurate balanced reports, as they should, but even the AP cannot vouch for the reliability of all of its sources. Interpreting information found in the mainstream press can be a challenge.

In environmental documents, you may occasionally encounter the wording "scientifically valid." For example:

EPA Action Item: Region 9 is providing oversight of an Arizona Department of Environmental Quality (ADEQ) environmental justice pilot project funded by a grant from Headquarters. The pilot project will be used to support Agency efforts to develop *scientifically valid* standards to measure cumulative risk. (Source: www.epa.gov/swerosps/ej/html-doc/health.htm)

or

Question: May HC-12a be used to replace CFC-12, commonly referred to as Freon, in cars?

Answer: HC-12a may not be used as a substitute for CFC-12 under any circumstances in an automobile. EPA has concerns about the safety of using a flammable refrigerant in a system not designed to reduce the risk posed by that flammability. EPA requires that a submitter of a flammable substitute conduct a *scientifically valid*, comprehensive risk assessment. To date, no material submitted by OZ Technology to EPA demonstrates the safety of using HC-12a in automobile air conditioning systems. (Source: www.epa.gov/spdpublc/title6/snap/oldhc12a.html)

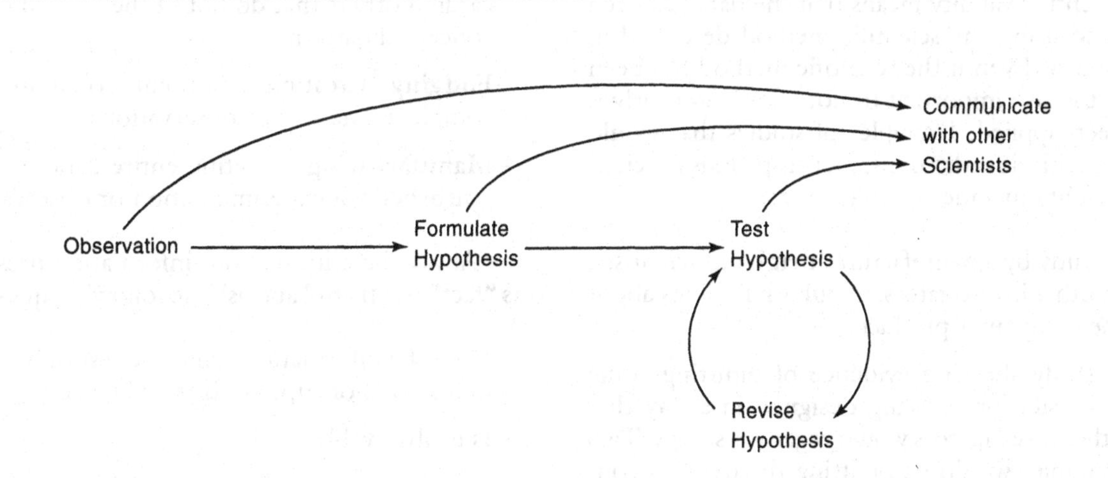


Since environmental science requires the analysis of scientific input, it is appropriate to understand how scientists gather and evaluate information....

The word *science* creates a variety of images in the mind. Some people feel that it is a powerful word and are threatened by it. Others are baffled by topics that are scientific and, therefore, have developed an unrealistic idea that scientists are brilliant individuals who can solve any problem that comes along. Neither of these images accurately reflects what science is really like. Science is a body of knowledge characterized by the requirement that information be gathered and evaluated by impartial testing of hypotheses and that information be shared so that it can be evaluated by others in the field. The scientific method of gathering information generally involves the following elements: observation, hypothesis formation, hypothesis testing, critical evaluation of results, and the publishing of findings.

Underlying all of these activities is constant attention to accuracy and freedom from bias. Observation simply means the ability to notice something. Sometimes, the observation is made with the unaided senses – we see, feel, or smell something. Often, special machines, such as microscopes, chemical analyzers, or radiation detectors, may be used to extend our senses. Because these machines are often complicated, we often get the feeling that science is incredibly complex, when in reality the scientific questions being asked might be relatively easy to understand.

When a question has been formulated and needs scientific investigation, the first step is the formation of a hypothesis. A hypothesis is a logical statement that explains an event or answers a question. A good hypothesis should be as simple as possible, while taking all of the known facts into account. Furthermore, a hypothesis must be testable. In other words, you must be able to support it or prove it incorrect. The construction and testing of hypotheses is one of the most difficult (creative) aspects of the scientific method. Often, artificial situations must be constructed to test hypotheses. These are called experiments. A standard kind of experiment is one called a controlled experiment. A controlled experiment is one in which two groups are created that are identical in all respects except one. The control group has nothing done to it that is out of the ordinary. The experimental group has one thing different. If the experimental group gives different results from the control group, it must be the result of the single difference (variable) between the two groups.

The results of a well-designed experiment should be able to support or disprove a hypothesis. However, this does not always occur. Sometimes, the results of an experiment are inconclusive. This means that a new experiment needs to be conducted or that more information needs to be collected. Often, it is necessary to have large amounts of information before a decision can be made about the validity of a hypothesis. The public often finds it difficult to understand why it is necessary to perform experiments with large numbers being involved, or why it is necessary to repeat experiments again and again.

The concept of repeatability is important to the scientific method. Because it is often not easy to eliminate unconscious bias, it is useful to have independent investigators repeat the same experiment to see if they get the same results. To do this, they must have a complete and accurate written document to work from. This process of publishing results for others to examine and criticize is one of the most important steps in the process of scientific discovery. If a hypothesis is supported by many experiments and by different investigators, it is considered reliable.

Enger, Eldon D. and Smith, Bradley F., *Environmental Science: A Study of Interrelationships*, 5th Edition, Dubuque, IA: Wm. C. Brown Publishers, 1995, p. 37. Reproduced with permission of The McGraw-Hill Companies.

Figure 2-6: Excerpt on Scientific Thinking from *Environmental Science: A Study of Interrelationships*, 5th Edition by Enger and Smith

Scientific validity means that the data has been obtained using the scientific method described in Figure 2-6. Even if the scientific method has been used, it is also important to note how rigorously it has been applied. Examples of studies that would make a critical reader or listener suspicious of scientific validity include the following:

1. A study by a manufacturer that hired scientists, or other investigators, to publish findings about the company's product.
2. A study showing evidence of omitting some basic step or of being designed in a way that other investigators would regard as sloppy. Two examples would be omitting the use of a control group or failure to use a statistically representative sample size.
3. A study that was not verified by other scientists – that is, it has not or cannot be replicated.

**Replicated:** Able to be repeated

**Statistically sound:** Having a firm basis in statistics; free from logic errors

**Verifiable:** Able to determine accuracy by investigation

The following are ways in which scientists can misrepresent scientific data and findings:[1]

—**Massaging** Performing extensive transformations or other maneuvers to make inconclusive data appear to be conclusive

—**Extrapolating** Developing curves based on too few data points, or predicting future trends based on unsupported assumptions about the degree of variability in factors measured

—**Smoothing** Discarding data points too far removed from expected or mean values

—**Slanting** Deliberately emphasizing or selecting certain trends in the data and ignoring or discarding others that do not fit the desired or preconceived pattern

—**Fudging** Creating data points to augment incomplete data sets or observations

—**Manufacturing** Creating entire data sets without benefit of experimentation or observation

Always be critical about information presented as "fact" or "truth" and ask the following questions:

—Does the information stand the test of being demonstrable or supported by evidence?

—Is it observable?

—Has the scientific method been applied properly? Might the information be an assumption?

In Figure 2-6, "Scientific Thinking" is excerpted from a textbook entitled *Environmental Science: A Study of Interrelationships*, by Eldon D. Enger and Bradley F. Smith. Enger and Smith's explanation of the components of the scientific method can help apply critical thinking skills to what we read and hear.

## Checking Your Understanding

### Activity A

Find a "science" article in an issue of tabloid newspaper that specializes in gossip. Analyze the article by marking the following:

—Examples of statements claiming to be facts or data

—References to "experts"

—References alleging to be reports by governments or research institutions

In one or two paragraphs, write whether you think that the tabloid article adopted a fantasy approach or tried to present a serious, factual tone; identify "facts" that are unsupported; and summarize your feelings about the value of tabloids. Be sure to include the article or a copy of the article with your paragraphs.

[1]: Sindermann, Carl J., *Winning the Games Scientists Play*, New York, Plenum Press, 1982, p. 193

## Activity B

Select a "science" article from a local newspaper or a publication such as *Newsweek* and another one from a tabloid. After carefully reading the two articles, write several paragraphs comparing/contrasting the differences in the way "factual" information was presented. Be sure to include the articles or copies of the articles with your analysis.

## Activity C

Clip or copy a journal or newspaper article that makes use of statistics and/or scientific data from the social, political, and basic science areas. Write a paragraph describing the evidence that you would need to regard the data as scientifically valid. Be sure to include the article or a copy of the article with your description.

## Activity D

After reading the following article, do you find the statistics referenced credible? Why or why not? Write a short explanation of your position.

### MTBE of Dubious Value in Reducing Air Pollution

Methyl t-butyl ether (MTBE) and other oxygenates have no significant effect on exhaust emissions from advanced-technology vehicles. According to a study commissioned by the State of California and conducted by researchers at the University of California, Davis, there is no significant difference in the emissions reduction of benzene between oxygenated and non-oxygenated California Phase II reformulated gasoline that meets all other standards. Thus, concluded the researchers, there is no significant additional air-quality benefits to the use of oxygenates such as MTBE in reformulated gasoline relative to alternative non-oxygenated formulations. The report says that MTBE should be phased out of use in California because the water contamination problems that have occurred since its introduction make its cost not worth the supposed air-quality benefits. However, fuel oxygenate content is mandated by federal law, so banning the fuels may not be a viable option. The Oxygenated Fuels Association responded that advanced-technology vehicles make up only about 32% of the vehicles in California and that nearly 70% of California's vehicles would continue to enjoy significant anti-pollution benefits from using cleaner burning fuel. The report is available at http://www.tsrtp.udcavis.edu/mtberpt.

# 2-5 Research: Checking Internet Sources

## Concepts

- The Internet is unregulated, and information obtained from it must be scrutinized.
- Indicators of credibility are the author, the publishing body, and the quality of the Internet site.

The user always has the burden of determining the validity of information on the Internet.

It is important to recognize that the Internet is free from government regulation and lacks any kind of quality control. In such an environment it is critical to remember that the burden of determining the validity of information found on the Internet is on the user. The basic questions for evaluating the validity of the information on the Internet are similar to those used in traditional library research. They include the following:

—Who wrote it?

—Who published it?

—What is the point of view?

—Why was it written?

—Are references made to other reputable sources?

Authorship, publishing body, and quality of the Internet site are major indicators of credibility. Authorship refers to the person who wrote the piece. Critical questions regarding authorship include:

—Is the author well known?

—Is the author well regarded in this particular field of study?

—Does the Internet site post a biographic sketch of the author, including the author's position, institutional affiliation, and address?

—If biographical information is not available, is there a phone number, mailing address, and e-mail address where additional information can be obtained?

—If biographical information is available, does the person appear qualified to write about the topic? (For example, a physician specializing in skin cancer would be a more credible source on the topic of skin cancer than the vice president of a large tanning bed manufacturer.)

If the answer to even a few of these questions is no, the information provided on the Internet site is suspect. The reader should be careful and locate substantiating documentation before assuming the credibility of the original document.

In the traditional world of print, the publishing body is the publisher of the document. Knowing the publisher gives some indication of the review process used in the preparation of the document. On the Internet, publishing body often refers to the institution or person posting the site or homepage, or the group that monitors the server.

The publishing body can give very good clues as to the review process used in the development of the information. Internet addresses ending in .edu and .gov indicate sites posted by educational institutions and governmental organizations. Reputable educational institutions, such as colleges and universities, monitor what is published under their name. This gives you some assurance that the material has been submitted to a review process. Government publications and, in many cases, large newspapers and professional journals also have a review process. Therefore, when the information is posted to a site monitored by a major university, a governmental entity (e.g., Environmental Protection Agency), a prestigious newspaper (e.g., *The Christian Science Monitor*) or a professional journal (e.g., *Journal of the American Medical Association*), a valid editorial process is highly likely. For specific information regarding these sites' review processes, contact the institutions directly.

Questions for evaluating publishing bodies include the following:

—Is the name of the organization posted on the document, in headers, footers, or a distinctive watermark that shows the document is part of a scholarly Web site?

—Is the organization national?

—Does the organization maintain its own server? (In general, information posted by an individual or supported by an Internet service provider is more suspect than information from a site that is maintained and owned by an organization. Self-support is an indication that the organization has solid financial backing and, consequently, the resources to publish more comprehensive and accurate information.)

—Is the organization recognizable and respected within the field?

—Is the organization known for being unbiased or funded by a group with a particular bias?

—Is the information posted to a site maintained by an organization free from any political or philosophical agenda?

—Is there a professional relationship between the author and the publishing body and was the document prepared as part of that individual's professional duties? (For example, if the site is a university, is the author a faculty or staff member at that institution?)

—Is the information published by a business? It is likely that the underlying motive for the site is to sell something or present the company in the best possible light?

A final consideration is the visual quality of the Internet site. Don't assume expensive looking sites provide accurate information. For example, many U. S. government sites have no graphics and are not structured very well, but the information is credible. In general, however, it is a good idea to ask the following questions about an Internet site before taking the information it contains seriously:

—Is the site well organized?

—Are words spelled correctly?

—Is the quality of the writing at the level you would expect from a good newspaper?

—Does the site download properly?

—Do the links work and, if so, do they link to other credible sites such as those maintained by universities and governmental entities?

—Is the material current and is the date the site was last maintained recent?

The questions throughout this section have all been posed in such a way that "yes" responses usually indicate credibility of a site, while "no" responses usually indicate that a site is suspect. Therefore, if you evaluate an Internet site using those questions, then a high number of yes responses indicates greater validity. A high number of no responses indicates less validity. More research will be required to determine credibility.

## Checking Your Understanding

### Activity A

Select a World Wide Web search engine (Infoseek, Yahoo, Excite, etc.), enter a topic (e.g., "environmental education" or "Earth Day"), and go to two related sites – one that appears somewhat suspect and another published by a credible university or government agency. Use the questions listed in this section regarding authorship, publishing body, and quality of Internet site to assess the validity of the sites. Write an analysis of the sites' credibility.

**Examples of Key Word Searches**

When searching for a phrase on the Internet, either enclose it between quotation marks or put a plus (+) between each word, so they will be considered together in the search. More detailed information about search techniques can be found by clicking on a help or search instructions link.

The following are examples of possible of key word searches:

HAZWOPER
"hazardous waste site worker"
or
hazardous + waste + site + worker
"hazardous waste operations"
"29 CFR 1910.120"
"40-hour training"

### Activity B

Using one of the previously mentioned search engines (Infoseek, Yahoo, Excite, etc.), search for two sites that provide Material Safety Data Sheet (MSDS) databases. Select a common chemical, such as sulfuric acid, $H_2SO_4$, and print out the MSDS for each site. Write a brief explanation of the similarities and differences you observe. Then write a paragraph explaining why it would or would not be permissible to use either of the MSDSs for worker training.

# 2-6 Outline: Building the Framework

## Concepts

—An outline is a useful organizing tool for writing and reading long and short documents.

—An outline can help you focus on key points and get the writing process started.

—Outlines can be alphanumeric or numeric.

You have clearly defined the central topic of your document and brainstormed all of the issues involved. You have consulted the CFR and read various codes, specifications, and industry standards. Sitting on your desk is a stack of research notes. Now what do you do? The next step is creating an outline. An outline is a valuable writing tool for

—Organizing research and ideas

—Identifying key points

—Summarizing the information

In general, an outline is used for longer documents. However, it can also be a helpful organizing tool for memos and letters (see Chapter 3). One way to begin an outline is to organize research notes and brainstorm topics that are similar to one another into groups. The next step is to write a sentence or phrase describing each group's similarity. Be sure that your descriptions are clear and concise. This will help to solidify your thoughts. After you are satisfied that all research notes and brainstorm topics have been grouped and labeled correctly, place them in a logical order, such as chronological order or level of importance.

Next, take a look at the headings and ask if they can stand by themselves. If a heading actually supports another heading, then make it a subheading. If the headings are logically ordered and comprehensive enough to base your document, letter, or memo on, then begin writing. If more structure and detail is needed, add subheadings.

The structure of an outline may be general, showing just key points and sub-points, or it may be quite specific, breaking sub-points into ever more detailed levels. The desired level of detail is a matter of personal preference. In general, a more detailed outline makes writing easier. Figure 2-7 shows an example of a traditional, alphanumeric outline. Figure 2-8 shows how more sublevels can be added to the outline shown in Figure 2-7.

For long documents, you may want to use your outline's headings and numeric system directly. The advantage of a numeric system is that the numbering links each detail backward to its main idea. This is particularly useful when the supporting text for a subheading covers multiple pages. Using an outline format for a document's headings also makes it easier for the reader to skim forward to the next topic.

Indentation is another useful visual aid that lets readers know what level of the outline they are on. The numeric outlines in Figure 2-9 and Figure 2-10 show the effects of indenting and not indenting.

In addition to being a writing tool, an outline is also a useful reading tool, especially for gaining a better understanding of long, complex documents. Figure 2-11 is an example of how an outline can be used to summarize a document. For simplicity, the example document is a short paragraph and the outline covers the two highest outline levels.

**Hazardous Waste Site Safety Procedures**

I. Investigating potentially hazardous atmospheres
  A. Permit-required confined spaces
    1. Air monitoring procedures
      a. Oxygen monitoring
        (1) Acceptable readings
        (2) Unacceptable readings
      b. Combustible gas monitoring
      c. Toxic gas monitoring
      d. Radiation monitoring
    2. Emergency response procedures
      a. Buddy system
      b. Communications components
      c. Safety lines and harnesses
    3. Documentation procedures
      a. The permit
      b. Training records
      c. Field log entries
  B. Flammable and explosive atmospheres
  C. Toxic atmospheres
  D. Radioactive atmospheres
II. Decontaminating equipment
  A. ...
  B. ...
III. Training hazardous waste site personnel
  A. ...
  B. ...

Figure 2-7: Alphanumeric Outline Example

**Hazardous Waste Site Safety Procedures**

I. Investigating potentially hazardous atmospheres
  A. Permit-required confined spaces
    1. Air monitoring procedures
      a. Oxygen monitoring
        (1) Acceptable readings
          (a) OSHA standard
            (1) Oxygen deficient atmosphere
              a) <19.5%
        (2) Unacceptable readings

Figure 2-8: Alphanumeric Outline Example with Additional Sublevels

**Title: Hazardous Waste Site Safety Procedures**

1 Investigating potentially hazardous atmospheres
  1.1 Permit-required confined spaces
    1.1.1 Air monitoring procedures
      1.1.1.1 Oxygen monitoring
        1.1.1.1.1 Acceptable readings
        1.1.1.1.2 Unacceptable readings
      1.1.1.2 Combustible gas monitoring
      1.1.1.3 Toxic gas monitoring
      1.1.1.4 Radiation monitoring
    1.1.2 Emergency response procedures
      1.1.2.1 Buddy system
      1.1.2.2 Communications components
      1.1.2.3 Safety lines and harnesses
    1.1.3 Documentation procedures
      1.1.3.1 The permit
      1.1.3.2 Training records
      1.1.3.3 Field log entries
  1.2 Emergency response procedures
    1.2.1 Buddy system

Figure 2-9: Numeric Outline Example with Indentation

**Title: Hazardous Waste Site Safety Procedures**

1 Investigating potentially hazardous atmospheres
1.1 Permit-required confined spaces
1.1.1 Air monitoring procedures
1.1.1.1 Oxygen monitoring
1.1.1.1.1 Acceptable readings
1.1.1.1.2 Unacceptable readings
1.1.1.2 Combustible gas monitoring
1.1.1.3 Toxic gas monitoring
1.1.1.4 Radiation monitoring
1.1.2 Emergency response procedures
1.1.2.1 Buddy system
1.1.2.2 Communications components
1.1.2.3 Safety lines and harnesses
1.1.3 Documentation procedures
1.1.3.1 The permit
1.1.3.2 Training records
1.1.3.3 Field log entries
1.2 Emergency response procedures
1.2.1 Buddy system

Figure 2-10: Numeric Outline Example without Indentation

**Paragraph**

Written procedures shall be prepared covering safe use of respirators in dangerous atmospheres that might be encountered in normal operations or in emergencies. Personnel shall be familiar with these procedures and the available respirators. In areas where the wearer, with failure of the respirator, could be overcome by a toxic or oxygen-deficient atmosphere, at least one additional man shall be present. Communications (visual, voice, or signal line) shall be maintained between both or all individuals present. Planning shall be such that one individual will be unaffected by any likely incident and have the proper rescue equipment to be able to assist the other(s) in case of emergency. When self-contained breathing apparatus or hose masks with blowers are used in atmospheres immediately dangerous to life or health, standby men must be present with suitable rescue equipment. Persons using air line respirators in atmospheres immediately hazardous to life or health shall be equipped with safety harnesses and safety lines for lifting or removing persons from hazardous atmospheres or other and equivalent provisions for the rescue of persons from hazardous atmospheres shall be used. A standby man or men with suitable self-contained breathing apparatus shall be at the nearest fresh air base for emergency rescue.

**Outline**

I. Written procedures for respirator safety
  A. Potential respirator failure
    1. In toxic or oxygen-deficient atmospheres
      a. Additional personnel required
      b. Communication requirements
      c. Rescue equipment available
    2. In "immediately dangerous to life or health" atmospheres
      a. Standby personnel required
      b. Suitable rescue equipment available
    3. In "immediately hazardous to life or health" atmospheres
      a. Safety harnesses and lifting lines required
      b. Standby personnel with SCBA required

Figure 2-11: Analysis of Paragraph Using an Outline

# Checking Your Understanding

## Activity A

Outline the article below by identifying the topic sentence and its supporting ideas.

**U.S. Signs Global Climate-Change Pact**

The U.S. has signed the Kyoto protocol, the United Nations treaty that will require industrialized nations to reduce their emissions of greenhouse gases to limit the potential for global climate change. The treaty was signed last Thursday by A. Peter Burleigh, deputy U.S. representative to the UN, at the UN in New York City. Signing the treaty was adamantly opposed by many members of Congress, who fear the mandatory reductions could harm the nation's economy. The action by the U.S. took place near the close of two weeks of negotiations on the pact in Buenos Aries. There, much debate centered on voluntary commitments to emissions reductions by developing nations, which currently have no mandate to do so under the Kyoto protocol. However, Argentina has committed itself to such reductions and, late last week, was trying to convince other developing nations to do the same. But the two largest developing countries, China and India, remained strongly opposed to setting emissions restrictions. Russia is now the only industrialized nation that has not signed the treaty.



## Activity B

Find an example of a technical outline, such as in a textbook or software user's manual. Write down what type of outline is being used and if the levels and sublevels are indented. Analyze the first section (main idea plus supporting details) and write a paragraph supporting your opinion about whether the formatting assists in the organization of the information or not.

## Activity C

As the environmental technician at Locust®, you are preparing for the many environmental health and safety issues facing you. You decide to list the issues in a simple outline format. Refer to Scenario 1 in Appendix 1 to fill in the following basic outline. If you have a background in environmental technology, you may add to the outline other issues that you would expect to address. The start of your outline may look something like the following:

I. Safety and Health Compliance Issues
   A.
   B.
   Etc.

II. Environmental Compliance Issues
   A.
   B.
   Etc.

# 2-7 Write: Getting Started

## Concepts

- When using an outline to write a document, use the outline for the document's heading and then write the details of the heading's subject.
- Lists and illustrations can facilitate faster and easier understanding of technical documents.

Sometimes the hardest part about writing is getting started. The good news is that if you have completed an outline as described in the previous section, you have already begun writing. The next step can be either of the following:

—Developing document headings from your outline

—Providing the details to substantiate the outline headings

Headings signal the reader about the document's content as well as how it is organized. There are two main considerations when developing headings: wording and style. Each contribute to the reader's understanding of the document's organization.

The wording of a heading prepares a reader for the upcoming paragraphs. When writing headings, check for the following:

—**Brevity** Headings should be as short as possible, using phrases (e.g., Handling Drums and Other Containers), single words (e.g., Training), or the briefest of sentences (e.g., What is a Site Emergency?).

—**Consistency** The word structure should be the same throughout each level. Use either naming word phrases (e.g., Decontamination or Site Characterization) or action phrases (e.g., Review Available Information). Phrases may end in "ing," (e.g., Reviewing Available Information).

—**Punctuation** Avoid punctuation, except for run-in headings, which can be followed by a colon or dash.

Style refers to choices such as bolding, underlining, italicizing, indenting, capitalizing, centering, line spacing around the header, and font (size and appearance of the typeface). Heading styles provide readers with a visual indicator of the level they are reading, allowing them to know where they are within the document structure. Each of the heading levels should have its own consistent style. One example might be as follows:

—Level 1 main ideas: numbering, large boldface type, upper and lower case

—Level 2 supporting ideas: smaller boldface type, upper and lower case

—Level 3 sublevels to the supporting ideas: boldface type, upper and lower case, appearing on the same line as the text that follows

Figure 2-12 shows the styles of headings used in this book.

| | |
|---|---|
| Level 1 Heading | **3-4 Memo** |
| Level 2 Heading | **Memo Formats**<br>Memoranda, referred to as memos, are letters sent to colleagues within the same company. There are many acceptable styles for these communications. In fact, businesses often adopt one style for company-wide use. In general, business memos consist of three parts . . . |
| Level 3 Heading | **Body of the Memo**<br>In general, a memo consists of an opening statement and a discussion. The opening statement usually communicates the purpose of the memo, some brief background information, and the specific course of action the reader is being asked to take. It should be clear and concise and . . . |

Figure 2-12: Heading Styles Orient the Reader

Paragraph text can be developed from outline headings as shown in Figure 2-13. How to write paragraphs to support an outline is the subject of many English textbooks and too broad a topic for the scope of this handbook. For such information, refer to the bibliography at the end of this book.

When writing supporting paragraphs for technical or business documents, be sure to use as many examples, lists, and illustrations as possible. Lists are used in technical documents frequently, and for good reason. Compare the two passages in Figure 2-14 regarding noise hazards. Which is easier to read and interpret?

Most people would agree that the second passage is easier to read than the first passage. Lists are easier to read and write than paragraphs. The passage, which is from NIOSH/OSHA/USCG/EPA's *Occupational Safety and Health Guidance Manual for Hazardous Waste Site Activities*, might be further improved by breaking out the details of the second paragraph as shown in Figure 2-15.

A list may be used for any content that includes a series of thoughts. The conventions regarding lists are as follows:

—Include an explanatory statement before a list. Avoid starting a list directly after a heading. The statement may be a complete sentence, such as the one preceding this list or a partial sentence that forms a complete thought when it is linked to the items below it.

—Use a colon (:) after the explanatory statement.

—Use periods after items that are sentences or lengthy phrases. Omit end punctuation if the items are briefly stated.

—Select numbered or alphabetized lists for items that are sequential.

**Developing a Paragraph from an Outline**

I. Identifying Hazardous Materials
   A. OSHA Container Requirements
      1. Hazard Communication Training
      2. Labels
         a. Size
         b. Color
      3. Markings
         a. Proper Chemical Name
         b. Name of Manufacturer
         c. Instructions/Cautions
   B. DOT Requirements
      1. Hazard Definition
      2. Hazard Classes
         a. Vehicle Placards
         b. Shipping Papers
   C. NFPA 704 System
      1. Hazard Code Diamond
         a. Health – Left Quadrant – Blue
         b. Fire Hazard – Top Quadrant – Red
         c. Reactive Hazard – Right Quadrant – Yellow
         d. Specific Hazard – Bottom Quadrant – White
      2. Hazard Rankings
         a. Extreme = 4
         b. High = 3
         c. Moderate = 2
         d. Slight = 1
         e. No = 0

Over the years, agencies have developed different regulations and requirements to alert users and emergency response personnel that containers, vehicles, or buildings contain hazardous materials. For example, OSHA regulations require hazardous communication training, warning labels, and markings that provide the proper name, manufacturer, and special precautions on containers of hazardous materials used by workers. In general, DOT defines a hazardous material as any substance that has been determined by the Secretary of Transportation to be capable of posing an unreasonable risk to health, safety, and property when transported in commerce. It uses nine hazard classifications and requires that diamond-shaped placards announcing the hazard be displayed on the front, back, and sides of all vehicles transporting hazardous materials. DOT regulations also require shippers to communicate the hazards of their cargo by providing shipping papers if hazardous material is sent and a Uniform Hazardous Waste Manifest if waste is sent. The NFPA 704 labeling system uses a color-coded diamond to identify the hazards associated with a particular chemical. Information about its health hazards (blue), flammability (red), reactivity (yellow), and other hazards (white) are identified in each of its sectors. A four-number ranking system is used for each sector with a 4 indicating the highest to a 0 indicating that no hazard is present.

Figure 2-13: Example of How a Paragraph Can Be Developed from an Outline

—If you use bullets, select a style and use it throughout a document. It is acceptable to use a second style of bullet to add a list of details under an item (as in Figure 2-16). If you do, be consistent. Too much variety can cause confusion for readers.

—Begin each item with a capital letter.

—Use parallel wording for items in a list. In other words, start each with an action word (as this list does) or with a naming word (as in Figure 2-15). Use all sentences (as this list does) or all phrases (as in Figure 2-16).

| Passage One | Passage Two |
| --- | --- |
| Work around large equipment often creates excessive noise. The effects of noise can include workers being startled, annoyed, or distracted. Excessive noise can also cause physical damage to the ear, pain, and temporary and/or permanent hearing loss. Furthermore, a noise can interfere with communication, which may increase potential hazards to employees due to the inability to warn of danger and proper safety precautions. If employees are subjected to noise exceeding an 8-hour, time-weighted average sound level of 90 dBA (decibels on the A-weighted scale), feasible administrative or engineering controls must be utilized. In addition, whenever employee noise exposures equal or exceed an 8-hour, time-weighted average sound level of 85 dBA, employers must administer a continuing, effective hearing conservation program as described in OSHA regulation 29 CFR Part 1910.95. | Work around large equipment often creates excessive noise. The effects of noise can include:<br><br>— Workers being startled, annoyed, or distracted.<br><br>— Physical damage to the ear, pain, and temporary and/or permanent hearing loss.<br><br>— Communication interference that may increase potential hazards to employees due to the inability to warn of danger and proper safety precautions to be taken<br><br>If employees are subjected to noise exceeding an 8-hour, time-weighted average sound level of 90 dBA (decibels on the A-weighted scale), feasible administrative or engineering controls must be utilized. In addition, whenever employee noise exposures equal or exceed an 8-hour, time-weighted average sound level of 85 dBA, employers must administer a continuing, effective hearing conservation program as described in OSHA regulation 29 CFR Part 1910.95. |

Figure 2-14: Example of Text Presented First in Paragraph and then in List Form

OSHA has regulations pertaining to employees who are subjected to excessive noise at work. Over an 8-hour shift, if the average* sound level equals or exceeds the following levels, the employer must take action:

— If readings exceed 90 dBA (decibels on the A-weighted scale), use feasible administrative or engineering controls.

— If readings equal or exceed 85 dBA, administer a continuing, effective hearing conservation program as described in 29 CFR Part 1910.95.

*Readings are taken hourly in an 8-hour shift and then averaged. The regulations refer to this as an "8-hour, time-weighted average."

Figure 2-15: Second Paragraph of Passage Two in Figure 2-14 Presented in List Form

**Safety Hazards**

Hazardous waste sites may contain numerous safety hazards, such as:

- Holes or ditches
- Sharp objects, such as
  - Nails
  - Metal shards
  - Broken glass
- Slippery surfaces
- Steep grades
- Uneven terrain
- Unstable surfaces, such as
  - Walls that may cave in
  - Flooring that may give way
- Precariously positioned objects that may fall

Figure 2-16: Bulleted List Example

In addition to lists, graphics make documents easier to read. Graphics include tables, charts, graphs, drawings, flow charts, and photographs. The current versions of computer word processing programs, such as Microsoft® Word™ and Wordperfect™, include the capabilities to produce graphics, which make your documents more readable, attractive, and professional-looking.

Illustrations using simple, representational shapes (see Figure 2-17) can communicate complex ideas quickly and more easily to a wider audience. Graphics are especially useful for readers to whom English is a second language.

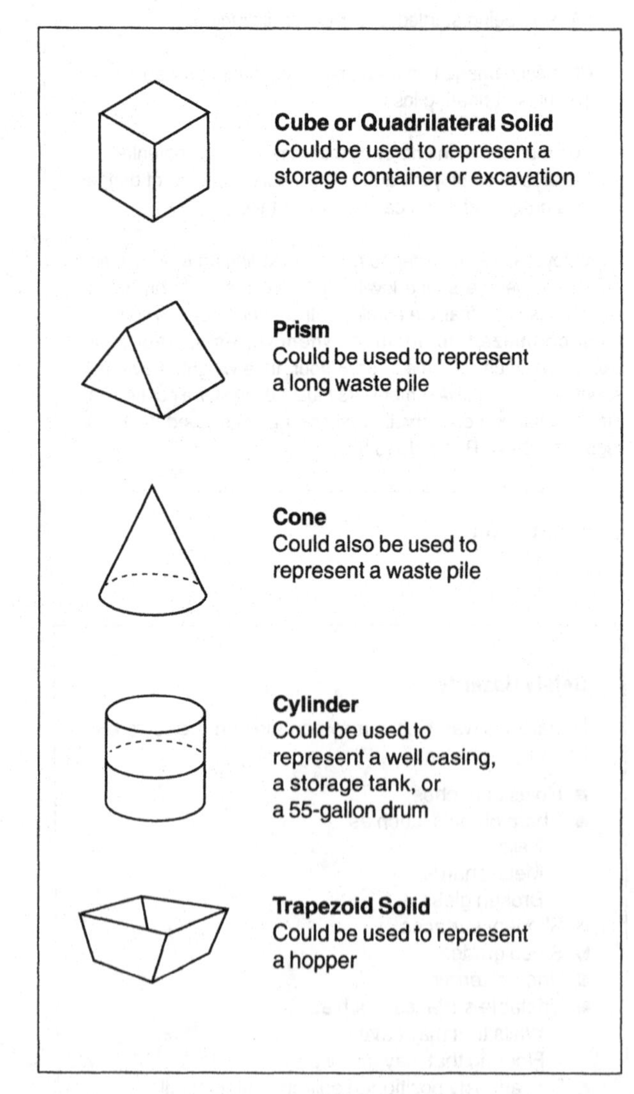


Figure 2-17: Basic Three-Dimensional Shapes

# Checking Your Understanding

## Activity A

Read Scenario 1 in Appendix 1 before starting this activity.

You are the new environmental technician at Locust®. Your boss wants you to begin building environmental safety and health awareness in the employees. You decide to start by writing about the use of appropriate eye and hearing protection. Use the following steps to write a memo:

1. Brainstorm a list of ideas that will cover the content of the memo.
2. Group the ideas and place them in a logical order.
3. Write a statement expressing the main idea of each grouping (i.e., a "topic sentence").
4. Apply to your groupings one of the outline formats described in the previous section.
5. If needed, add information to support the main ideas and supporting ideas. Insert the information into your outline.
6. Use headings and lists to develop the memo from your outline.

## Activity B

Illustrate how employees should use a Hazardous Material waste receptacle using geometric shapes to represent objects such as the receptacle, cans containing toxic materials, toxic liquids, etc. Use arrows, lines, and labels as necessary.

# 2-8 Edit and Proofread: Improving Documents

## Concepts

- All technical and business documents should be error-free.
- Editing and proofreading are part of the writing process.
- An editing checklist helps to ensure better, more consistent writing.

While it may not be very efficient to turn every document into a literary masterpiece, they should all be clearly written with no spelling, grammatical, or factual errors – especially if one misunderstanding by the user might result in a serious injury. Therefore, whenever you write always edit and proofread your work before it is sent out. Editing involves improving and clarifying language, and proofreading is checking for spelling, grammar, consistency, and accuracy of facts.

The following are several editing and proofreading tips. Select the ones that you can do and be diligent about using them every time you write.

1. Read your document aloud to yourself and identify any awkward sentence structures. Listen for incomplete thoughts (called sentence fragments) or thoughts that run together or that do not belong together.
2. If you write with a computer word processing program, use the spelling and grammar checking features. Before sending documents out, print a copy and check it again. It is easy to miss errors on the computer monitor. Note also that spell checking programs cannot detect the incorrect usage of words such as no and know, see and sea, there and their, etc.
3. Ask someone in your company who is good at editing and proofreading to review your work.
4. Use a checklist of common errors, such as the one below or those on the Internet. To find an Internet site, run a search using the keywords *proofreading* and *editing checklist*. The Yahoo search engine, for example, yields several outstanding checklists where you may find information by clicking on a topic (e.g., "Subject/Verb Agreement").

Figure 2-20 provides an example of a basic editing checklist.

| Common Usage Errors | |
|---|---|
| **Incorrect** | **Correct** |
| We was working.<br>They was working.<br>You was working. | We were working.<br>They were working.<br>You were working.<br>He, she, or it was working. |
| He don't mind helping.<br>She don't mind helping. | He (or she) doesn't mind helping.<br>They (or we) don't mind helping. |
| It don't matter. | It doesn't matter. |
| He done the work.<br>She done the work. | He (or she) has done the work.<br>He (or she) did the work. |
| She (or he) seen him. | She (or he) has seen him.<br>She (or he) saw him. |
| They (or we or you) seen him. | They (or we / you) have seen him.<br>They (or we / you) saw him. |

Figure 2-18: Common Usage Errors

| Basic Grammar Terms | |
|---|---|
| **Noun:** | Naming word, such as girl, room, or boredom. |
| **Plural:** | More than one. |
| **Singular:** | One |
| **Pronoun:** | Words such as he. she, it, we, you, they, I, and me, which substitute for a naming word, such as Mary, girl, or boss. |
| **Tense:** | The time of action in verbs, such as past, present, or future. |
| **Verb:** | 1) The action in a sentence, such as "He discovers the problem."<br>2) A limited group of words (am, is, are, was, were, be, been, seems, becomes, and remains) that show one's "state of being." For example, "He is calm." |

Figure 2-19: Basic Grammar Terms

## EDITING CHECKLIST

### Paragraph Organization

✓ check each item as you complete it

- ☐ Are paragraphs placed in an illogical order?
- ☐ Is there a topic sentence at the beginning of each paragraph?
- ☐ Do paragraphs' details support their topic sentences?

### Heading and List Consistency

| Incorrect | Correct |
|---|---|
| ☐ Is there an inconsistent use of style (i.e., fonts, capitalization, boldface, surrounding line spacing) within heading levels? | |
| **Heading 1**<br>Heading 1<br>*Heading 1* | **Heading 1**<br>**Heading 1**<br>Heading 2<br>Heading 2<br>*Heading 3*<br>*Heading 3* |
| ☐ Is there an inconsistent use of bullets or number styles within lists? | |
| ■ Item 1<br>● Item 2<br>○ Item 3 | ● Item 1<br>● Item 2<br>● Item 3 |
| ☐ Is there inconsistent wording with in lists? | |
| How to go to sleep:<br>1. Count sheep<br>2. Relaxing<br>3. Read a book | How to go to sleep:<br>1. Count sheep<br>2. Relax<br>3. Read a book |

### Sentence Structure

| Incorrect | Correct |
|---|---|
| ☐ Are there fragmented sentences (i.e., a sentence that does not express a complete thought? | |
| Which is attributed to global warming. | The ice in Antarctica is receding, which is attributed to global warming. |
| ☐ Are there run-on sentences (i.e., two sentences punctuated as one)? | |
| The EPA and OSHA developed the HAZWOPER standard, it is found in 29 CFR 1910.12.0. | The EPA and OSHA developed the HAZWOPER standard. It is found in 29 CFR 1910.120.<br><br>The EPA and OSHA developed the HAZWOPER standard, which is found in 29 CFR 1910.120. |
| ☐ Are there subject-verb agreement errors (i.e., a plural noun matched with a singular verb, and vice versa)? | |
| We is going.<br>She are going. | We are going.<br>She is going. |
| ☐ Are there compond pronoun errors? Note that a way to test for incorrect pronoun usage is to try one pronoun at a time with the rest of the sentence. In the example, "me designed the training" and "her designed the training" are incorrect. | |
| Me and her designed the training. | She and I designed the training. |

Figure 2-20: Example of an Editing Checklist

| **EDITING CHECKLIST** (continued) | |
|---|---|
| **Incorrect** | **Correct** |
| **Sentence Structure** (continued) | |
| ❑ Is tense usage inconsistent? | |
| We walked along the line as we observed the process and took notes. Suddenly, the line worker dropped a wrench. We *see* a spark and then everything*explodes!* | We walked along the line as we observed the process and took notes. Suddenly, the line worker dropped a wrench. We *saw* a spark and then everything *exploded!* |
| ❑ Are double negatives used? | |
| It didn't cause no problems. | It didn't cause any problems. |
| **Tone (Word Choice)** | |
| ❑ Is jargon used when readers are not familiar with it? | |
| (for non-business audience): We're going to do demographics before ramping up product. | (for non-business audience): We're going to conduct surveys before developing a new product. |
| ❑ Is slang used? | |
| I ain't buying it. It's a rip-off. | I'm not buying it. It's overpriced. |
| ❑ Are cliches or overused expressions used? | |
| It goes without saying that cleaning up that spill is easier said than done. | Cleaning up that spill will be difficult. |
| ❑ Are statements exaggerated? | |
| Our company is more than qualified to help you correct this extremely important situation. | Our company is qualified to help you correct this situation. |
| ❑ Are words used repetitiously? | |
| It is our consensus of opinion that the training class should cover the basic essentials of hazard recognition, so that a prompt and speedy response to unpredicted emergency events will be rapidly expedited. | It is our opinion that the training should cover the basics of hazard recognition to enable prompt responses to emergencies. |
| ❑ Is offensive language used? | |
| For example: Sexist or racist words. | |
| ❑ Is sarcasm used? | |
| For example: Nice going. Your accident-free record just bit the dust. | |
| ❑ Are accusations made? | |
| For example: You neglected to send me the name of the certified driller when I asked you to. | |

Figure 2-20 Continued

## Checking Your Understanding

### Activity A

Using the tips and checklists provided in this section, revise the letter below. If you require more than one attempt, staple all of the copies together with the final draft in front.

Dear Boss:

This here written letter is to warn you about potental problems in the nu program that give business credits for early action in an attempt to reduce greenhouse gas emissions from a companie. According to the reginal EPA guy, any reported emision reduction must be virified by an outside parti. It is crucil to 1st establish if the owner or operator of the business are responsible, elsewise there's the potential for a double reporting of the gas reduction. The 2nd problem involves determin a baselines form which to measure the emission reduction. Finally, the guy said it is important four companys to report emissions reduction on an organizationwide bases. Reporten on just a project could make it appear that a re-duction on emissions has occurred eventhough the companys' net total emissions may actually have inkreased.

Carl B. Careless

### Activity B

Trade your final version of the Activity A letter with another classmate. Edit and proofread each other's work.

## Summary

As an environmental technician you will be asked to write a variety of documents, as well as letters and memos. For all of these, the basic writing process is as follows:

1. Researching and gathering information
2. Outlining the document's structure
3. Writing the document
4. Editing and proofing

The main research resources you will use are the CFR, specifications, standards, codes, and the Internet. As you find information, carefully consider the scientific validity of the facts presented to you. Some things to watch for are authorship, publishing body, bias, motivation, referral to and/or knowledge of other sources, accuracy, and currency. Always check copyright permissions.

When you are ready to begin writing, use an outline to organize and solidify your thoughts. As you write, keep your headings consistent and use lists and illustrations wherever possible. When finished, always edit and proof your work, preferably with a checklist.

# 3

# Letters and Memos

## Chapter Objectives

Upon completing this chapter, the student will be able to:

1. **Demonstrate** the skills for writing the following correspondence specific to the environmental technician occupation:
   - Request, reply, and referral letters
   - Instruction and report letters
2. **Demonstrate** the skills for writing the following general business correspondence:
   - Confirmation, follow-up, and transmittal letters
   - Good will, support, recommendation, and thank-you letters
   - Claim, complaint, adjustment, and apology letters
   - Sales and collection letters
3. **Demonstrate** the skills for writing memos.

## Chapter Sections

# 3-1 Business Letter Formats

## Concepts

- On the job, environmental technicians are expected to write business letters.
- Business letters must follow a format that includes a letterhead, date, inside address, reference, salutation, body of text, closing, signature, and notations.
- Purpose and audience usually determine the tone of a letter.
- Business letters should be brief and to the point.
- The body of a letter should include an introduction with an opening statement, topic sentence, and preview of the letter; paragraphs explaining individual ideas; and a conclusion with a summary statement.

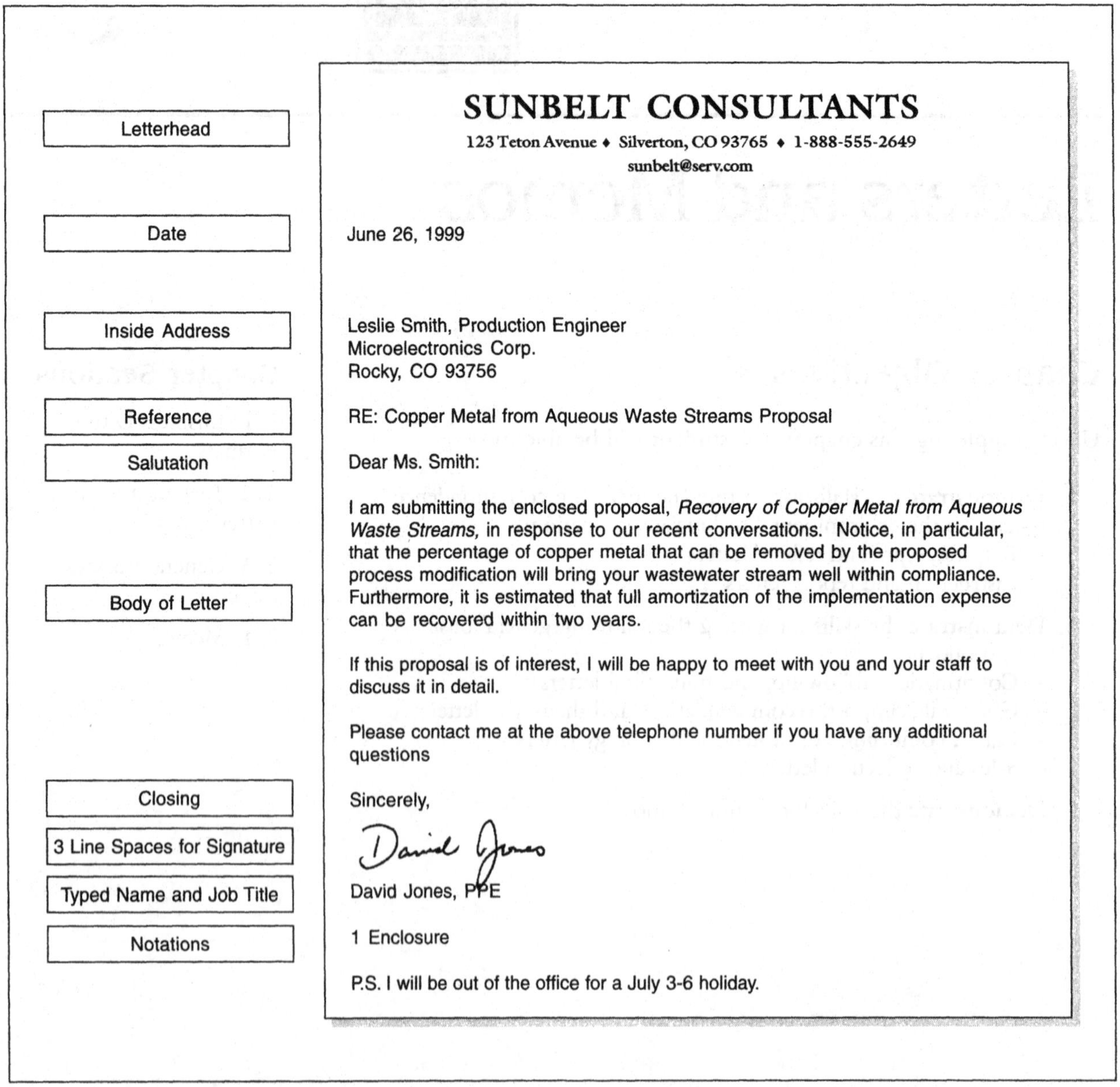

|  |  |
|---|---|
| Letterhead | **SUNBELT CONSULTANTS**<br>123 Teton Avenue ♦ Silverton, CO 93765 ♦ 1-888-555-2649<br>sunbelt@serv.com |
| Date | June 26, 1999 |
| Inside Address | Leslie Smith, Production Engineer<br>Microelectronics Corp.<br>Rocky, CO 93756 |
| Reference | RE: Copper Metal from Aqueous Waste Streams Proposal |
| Salutation | Dear Ms. Smith: |
| Body of Letter | I am submitting the enclosed proposal, *Recovery of Copper Metal from Aqueous Waste Streams,* in response to our recent conversations. Notice, in particular, that the percentage of copper metal that can be removed by the proposed process modification will bring your wastewater stream well within compliance. Furthermore, it is estimated that full amortization of the implementation expense can be recovered within two years.<br><br>If this proposal is of interest, I will be happy to meet with you and your staff to discuss it in detail.<br><br>Please contact me at the above telephone number if you have any additional questions |
| Closing | Sincerely, |
| 3 Line Spaces for Signature | David Jones |
| Typed Name and Job Title | David Jones, PPE |
| Notations | 1 Enclosure<br><br>P.S. I will be out of the office for a July 3-6 holiday. |

Figure 3-1: Basic Elements of a Business Letter

As an environmental technician you will have to write general business letters and letters specific to your occupation such as requests for material safety data sheets (MSDS) from manufacturers. Regardless of the content, your letters should follow the standard business letter format shown in Figure 3-1. This format consists of the following:

—**Letterhead** The letterhead is preprinted or printed on the stationery and contains the company name, address, telephone number, FAX number, and e-mail address.

—**Date** The date of writing is placed two lines below traditional letterheads or at the top margin if the letterhead is placed elsewhere on the stationery.

—**Address** The inside address is placed two to four lines below the date. The first line includes the recipient's name and business title (optional). The second line contains the company name. The third and fourth lines consist of the recipient's address, including zip code.

—**Reference (Optional)** The reference is placed two lines below the address. The subject line highlights what the letter is about. Use concise wording that gives the reader an overall idea of the contents of the letter, but don't be too general. For example, the subject "Contingency Plan" could mean anything from the announcement of a new contingency plan, to a commentary on problems with the current plan. A better subject line is "1998/1999 Contingency Plan Revision."

—**Salutation** The greeting is placed two lines below the reference or address. Greetings usually are written as one of the following:

- If you know the recipient, write Dear plus the first name followed by a colon (e.g., Dear John:).
- If you are not personally acquainted, write the last name with the title. Use Ms. for a woman's title and Mr. for a man's title (e.g., Dear Mr. Smith:).
- If you do not have access to names, write Dear Sir or Madam:, Dear Ladies and Gentleman:, or the reader's business title (e.g., Dear Personnel Manager:).

—**Body** The body, which is the letter's content, begins two lines below the greeting.

—**Closing** The closing is two lines below the body. Capitalize the first letter of the closing and end with a comma. Some traditional closings are Sincerely, Yours truly, Respectfully, and Best regards.

—**Signature** The signature area, which is four lines below the closing, includes your full name without Ms. or Mr. You may add your job title on the next line. Write your signature between the closing and your typed name.

—**Notations (Optional)** The notations area includes several possibilities:

- The initials (lower case) of the typist, if it was not you.
- The word Enclosure if there is one included in the mailing. If there are several enclosures, type the number and Enclosures (e.g., 4 Enclosures) or Enclosures followed by a list of their titles
- A message such as Certified, Delivery by messenger, or Under separate cover (meaning sent in a separate package) to document any special handling of the letter.
- The acronym cc: (carbon copy), or the phrase Copies to: followed by the names of others receiving the letter. This line is used if you want to send copies of the letter to other interested persons and to make the persons to whom the memo is addressed aware that others are reading the memo. The copy line is an efficient communication tool for informing individuals who are not directly involved with the subject of a letter, but still interested in learning about it. For example, copying a supervisor on letters regarding a particular project is an excellent way to keep him apprised of the progress you are making. Note that when individuals are copied on a letter they are just being informed of a situation and not being asked to take any specific action.
- The acronym bc: (blind copy) Blind copies are copies sent to additional people without the recipient's knowledge.
- The initials P.S. (postscript) to call attention to an added message.

The first step in writing a business letter is analyzing the audience and the purpose of the letter. The combination of purpose and audience analysis usually determines the tone (see Chapter 1) of a letter.

The next step in writing a business letter is organizing your thoughts into an outline. Most business letters follow the following outline:

I. Introduction
   A. Opening attention-getting statement(s)
   B. Topic sentence(s) stating reason for letter
   C. Preview of the contents of the letter
   D. Transition statement
II. Major Ideas
   A. First major idea of the letter in one sentence
      1. Explain the idea
      2. Support the idea
      3. Conclude the idea and provide a transition statement
   B. Sentence stating second major idea
      1. Explain the idea
      2. Support the idea
      3. Conclude the idea and provide a transition statement
   C. Sentence stating third major idea
III. Conclusion
   A. Summarize major ideas presented unless the letter is simple and short
   B. Make closing statement(s)

Note that you don't have to use all the above components, just those that are appropriate. An example of how the outline may be used to develop a letter follows. It is a sales letter from a hazardous waste cleanup company. The purpose of the letter is to introduce a new waste oil removal service to a commercial audience.

I. Introduction
   A. Opening attention-getting statement(s)
      *Are you spending more money than you should, disposing of your waste oil?*
   B. Topic sentence(s) stating reason for letter
      *I'd like to tell you about our newest waste oil recycling service, which will reduce your costs and liability.*
   C. Preview of the contents of the letter
      *I'll explain how your company can benefit from our entire service package.*
   D. Transition statement
      *Since keeping an eye on the bottom line is vital to a successful business, I'll begin by providing information on how you can reduce costs.*
II. Major Ideas
   A. First major idea of the letter in one sentence
      *We save you money because our costs for regular pick up and recycling of waste oil, including lube oil and engine oil, are the lowest in Southern California.*
      1. Explain the idea
         *Since our modern trucks make scheduled monthly pick-ups for thousands of customers, you realize the savings of not having to pay expensive individual pick-up rates where customers are charged additional mileage.*
      2. Support the idea
         *Our base of over 5,000 customers in the Southern California area using our monthly pick-up service significantly reduces the average transportation cost to each customer.*
      3. Conclude the idea and provide a transition statement
         *The same economy realized by making regular pick-ups also allows Waste Away Services to lower your liability.*
   B. Second major idea of the letter in one sentence
      *Regular pick-ups can lower your liability by reducing the storage time, which, in turn, reduces the likelihood of a costly spill.*
      1. Explain the idea
         *Allowing containers of waste oil to accumulate until an unscheduled pick-up can be made increases the probability that an accidental spill can occur.*
      2. Support the idea
         *The storage of waste containers is not only unsightly, but also exposes business owners to the additional financial risks of fires and vandalism.*
      3. Conclude the idea and provide a transition statement
         *The financial risks of allowing the accumulation of waste oil containers on your property can be catastrophic if a spill results in environmental contamination.*

1115 LaBrea Street
Angel City, California 90211

March 3, 1999

Don & Phil's Garage
5555 South Street
Everly, California 90807

Gentlemen:

Are you spending more money than you should for disposing of your waste oil? I'd like to tell you about our newest waste oil recycling service, which will reduce your costs and liability. I'll explain how your company can benefit from our entire service package. Since keeping an eye on the bottom line is vital to a successful business, I'll begin by providing information on how you can reduce costs.

We save you money because our costs for regular pick up and recycling of waste oil, including lube oil and engine oil, are the lowest in Southern California. Since our modern trucks make scheduled monthly pick-ups for thousands of customers, you realize the savings of not having to pay expensive individual pick up rates where customers are charged additional mileage. Our base of over 5,000 customers in the Southern California area using our monthly pick-up service significantly reduces the average transportation cost to each customer. The same economy realized by making regular pick-ups also allows Waste Away Services to lower your liability.

Regular pick-ups lower your liability by reducing the storage time, which, in turn, reduces the likelihood of a costly spill. Allowing containers of waste oil to accumulate for an unscheduled pick-up increases the probability that an accidental spill will occur. The storage of waste containers is not only unsightly, but also exposes business owners to the additional financial risks of fires and vandalism. The financial risks of allowing the accumulation of waste oil containers on your property can become catastrophic if a spill results in environmental contamination.

You can rid your business of unsightly waste storage containers, while at the same time reducing your liability by using our regularly scheduled pick-up service. Make the phone call today, so one of our knowledgeable sales technicians can analyze your waste pick-up needs, discuss our reasonable service fees, and place you on one of our regular pick-up schedules.

Sincerely,

Suzie Little
Waste Management Specialist

Figure 3-2: Sales Letter Developed by Using an Outline

III. Conclusion
- A. Summarize major ideas presented unless the letter is simple and short
  *You can rid your business of unsightly waste storage containers, while at the same time reducing your liability by using our regularly scheduled pick-up service.*
- B. Make closing statement(s)
  *Make the phone call today, so one of our knowledgeable sales technicians can analyze your waste pick-up needs, discuss our reasonable service fees, and place you on one of our regular pick-up schedules.*

When these separate sentences are put together, the body of the completed letter is illustrated in Figure 3-2.

# Checking Your Understanding

## Activity A

Select a non-personal business letter from your junk mail. Mark and identify each part of the letter's format as shown in Figure 3-1.

## Activity B

Using the same business letter you used in Activity A and the basic business letter outline in this section, make a list correlating the various outline topics (i.e., opening statement) with the appropriate sentences(s) in the letter.

# 3-2 Technical Business Letters

### Concepts

- Technical business letters are used for seeking or informing others of environmental information within a business setting.
- Technical business letters include request, reply, referral, report, and instruction letters.

## Request, Reply, and Referral Letters

These types of letters are used by environmental technicians to obtain and pass on technical information.

### Request Letters

**Purpose** To ask for information or a service.

**Contents** Request letters should include the following:

—Who you are

—What information or service is needed

—Why it is needed

—When it is needed

—How you can be reached for further information

—A thank you

**Comments** Environmental technicians typically use this type of letter for obtaining a Material Safety Data Sheet (MSDS). A request can be as simple as a brief letter that orders something (e.g., "Please send me your latest Hazardous Materials Safety Products catalog. My address is .... Thank you.") or a more complex request (see Figure 3-3 and Figure 3-4) asking for the reader's consideration of and action on something.

If you do not receive a response from an inquiry, it may be necessary to send a follow-up letter. See Figure 3-10.

### Reply Letters

**Purpose** To respond to a request.

**Contents** Reply letters should include the following:

—A reference to the reader's request

—A response to the request

—How you can be reached for further information

**Comments** If the information requested is not available or is confidential, simply explain why it can not be provided. If you know of a related source for the information, suggest that as an alternative. See Figure 3-5.

### Referral Letters

**Purpose** To direct the reader to another resource for information or assistance.

**Contents** Referral letters should include the following:

—A reference to the reader's request

—The name of the resource

—(Optional) Background information about the resource

—How to contact the resource

—How you can be reached for further information

**Comments** If the referral is in response to a request for information, the letter may be brief. For example, "In response to your question about where to get access to the CFRs, the reference room of the public library maintains a complete, up-to-date set."

Sometimes it is necessary for a referral letter to provide background information. See Figure 3-6.

## Checking Your Understanding

### Activity A

Write a letter to a manufacturer (Slick Oil) requesting an MSDS for one of their products (10-40 Oil) stating that you will need the sheet by the 15th of the next month.

### Activity B

As the previous owner of Prospector's Corner, write a reply to the request letter in Figure 3-3. You may be creative.

ENVIRONMENTAL CONSULTING SERVICE

100 Sockeye Road ■ Salmon, Idaho 83467 ■ 208-555-0001 ■ fwilliams@ecs.com

April 2, 1999

Ms. Paula Morgan
25 Scenic Drive
Mountain Home, Idaho 83647

Dear Ms. Morgan:

I am researching the history of a Salmon, Idaho, property referred to as Prospector's Corner. County records indicate that you owned the property from 1960 to 1975. Two neighbors recall that you removed an outbuilding to build your carpet store, which is still standing. One of the neighbors said that the outbuilding was used for a small chrome plating operation before you purchased the property.

Before a soil-sampling project on the site starts next month, I would like to gather as much historical data about the site as possible. Do you have any photographs, maps, or other descriptive materials regarding the site as it was when you purchased it? If you could recall the location of the outbuilding, it would be most helpful.

If you have questions about my request, I am available in the office daily between 3:00 p.m. and 5:30 p.m. My phone number, address, and e-mail address are on the letterhead. Thank you for your assistance.

Sincerely,

Fred Williams
Field Technician

Figure 3-3: Request Letter Example 1

SMOOTH COATINGS, INC.

June 1, 1999

Mr. John James
BioStuff Chemical Company
3501 Industry Road
Maple Valley, Ohio 44311

Dear Mr. James:

As the Environmental Health & Safety Officer for Smooth Coatings, it is my responsibility to review all Material Safety Data Sheets (MSDSs). The purpose of this letter is to inform you of an omission on two of your company's MSDSs. The products are:

1. Biocoat
2. Biocide Solvent

Since our Hazard Communication training is already scheduled, a response within the next five working days would be appreciated. If you need to contact me, please call 1-502-555-2002, extension 444. A quick response will be greatly appreciated.

Yours truly,

Al Jackson
Environmental Health & Safety Officer

925 Acorn Road, Lone Oak, Kentucky 42002 (502) 555-2002

Figure 3-4: Request Letter Example 2

3501 Industry Road
Maple Valley, Ohio 44311
(330) 555-4311

February 12, 1999

Al Jackson
Environmental Health & Safety Officer
Smooth Coatings, Inc.
925 Acorn Road
Lone Oak, Kentucky 42002

Dear Mr. Jackson:

Thank you for calling to our attention the missing sections on Biocoat's and Biocide Solvent's Material Safety Data Sheets. This past year we have been scanning our product's MSDS into a database so they can become available on our new web site (www.biochem.com/msds).

The products with the missing sections on their MSDS apparently occurred during the scanning process. We appreciate you calling these omissions to our attention and regret any inconvenience that may have resulted.

The complete MSDS for the two products will be sent to you by overnight mail. In the future, we anticipate that our valued customers will be able to obtain up-to-date Material Safety Data Sheets for all of our products from our Web site.

Sincerely,

Leslie Gray
Environmental Technician

Figure 3-5: Reply Letter

O'Donnell & Maxwell

Environmental Engineers

September 22, 1998

Lee Weatherby, Director
Skybart Project
2300 Civic Center Drive
Hanahan, South Carolina 29410

Dear Lee:

Thank you for the opportunity to serve you in investigating the Skybart site. I think we will all agree that this site has a tremendous potential for future development. As my enclosed report will outline, however, the preliminary investigation shows that the property has several areas of contamination that will need clean-up before the site can be marketed.

I would suggest that Sue Looshun at the EPA Region IV office would be a tremendous help with this clean-up effort. As the staff person who works with brownfields (abandoned industrial properties), she has a reputation for being a great resource. She can help the stakeholders to understand risk assessment and analyze their options, including how to fund the clean-up efforts. I think you will be very pleased with the assistance that Sue can provide.

It has been a pleasure working with the city staff. I hope to have the opportunity to serve you again in the future.

Yours truly,

Stan Baker
Waste Management Specialist

Enclosure

53 Hummingbird Lane, Pinehaven, South Carolina 29405 (843) 555-2942

Figure 3-6: Referral Letter

# ABC LAB SUPPLIERS

5115 Arrow Highway, Fargo, North Dakota 58109 (701) 555-7010

March 15, 1999

Mr. Mel Franz
ABC Laboratories
111 Meadow Road
Prairie Rose, North Dakota 58104

Dear Mr. Franz:

We are pleased that you have selected our Model 14 Gas Chromatograph to meet your laboratory's analysis needs. We regret that the operating manual was unavailable at the time of shipment and that you are having trouble regulating the flow of the carrier gas. Today, an order for your manual was placed with our Roanoke warehouse, and it should arrive by mail early next week.

Meanwhile, the following directions are being provided to assist you in calibrating the carrier gas flow to the instrument. These are the steps to be followed:

1. Check that all tubing connections have been made.
2. Open the shut-off valve on the helium tank and adjust the pressure regulator to between 7-10 $lbs/in^2$.
3. Prepare a detergent solution and place it in the rubber bulb attached to the lower end of the gas burette.
4. Attach the gas burette to the instrument's gas outlet port using a piece of small-diameter tubing.
5. Squeeze the rubber bulb forming a soap bubble just above the entry point of the gas into the burette.
6. Use a stopwatch to time the number of seconds it takes for the soap bubble to sweep the 50-ml volume.
7. Repeat the measurement at least three times and determine the average.
8. Calculate the flow of the carrier gas in ml/sec. If the flow rate is higher or lower than specified by the analytical procedure, adjust the pressure regulator accordingly and repeat Steps 4-7.

If for any reason you should continue experiencing a problem in obtaining the proper gas flow, please give our local sales representative a call, and he will come to your laboratory. We appreciate your business and again apologize for the inconvenience.

Sincerely,

Jake Arnold
Laboratory Technician

Figure 3-7: Instruction Letter

**AquaPure Systems**
333 Sweetwater Road, Wichita, Kansas 67276 • 316-555-6776

November 30, 1998

Amy Perkins, Regional Manager
AquaPure Systems
678 Highland Street
Kansas City, Missouri 64148

Dear Ms. Perkins:

I recently visited Bright Metal Corporation (BMC) at the request of our field manager Fred Neighbors, who requested that I report my findings directly to you. Bill Lever, BMC's Plant Manager, spoke to me about their wastewater treatment operation. BMC is a medium-sized plating shop that operates copper, bright chrome, and zinc plating lines. Bill told me that they have been having some problems meeting their discharge permit requirements and would be interested in learning more about the use of a reverse osmosis system.

Background information on BMC's operation is as follows:

1. All cyanide wastewater from the copper plating line is destroyed using the industry standard treatment with sodium hypochlorite at high pH to prevent the formation of deadly hydrogen cyanide gas.
2. After the cyanide destruction process is complete, this high pH waste stream is blended with the low pH waste stream from the chrome plating line.
3. The combined waste streams are then treated with a flocculant to coagulate the metal ions and assist in the formation of the metal hydroxide precipitate.
4. The outflow from the flocculating tank enters a clarifier to complete the separation of the floc from the clear effluent.
5. The floc is pumped from the bottom of the clarifier into a pneumatic filter press, where the excess water is removed.

The general findings of my visit are that they continue to have problems meeting their NPDES permit. I suggested that they consider the use of reverse osmosis, which would not only allow them to meet their permit requirements but provide them with a source of high-quality makeup water for their counterflow rinse system.

Mr. Lever is most interested in having one of our technical sales representatives visit their operation. I suggested that after an analysis of their waste stream, the proper unit could be recommended and, from the water savings realized, the amortization of the installation costs determined. I believe that Bill was pleased with our preliminary discussion. If the technical sales representative would like for me to accompany him, I will be happy to do so. If there is any other information that I can provide, please contact me.

Sincerely,

Sydney Green
Chemical Technician

cc: Fred Neighbors

Figure 3-8: Report Letter

### Activity C

You have been hired as a technician by Environmental Perfection Investigation and Consulting (EPIC). (See Scenario 2 in Appendix 1.) UDL asks you to respond to a letter from a customer inquiring about the Boom Box audio component of the integrated automobile dashboard. They would like to know the frequency range of the speakers. Write a referral letter to the customer providing him with the manufacturer's name, address, and phone number. Provide the name of a person or department in that company that can answer his questions. You can make these up.

## Instruction and Report Letters

These types of letters are used by environmental technicians for providing brief technical instructions and reports.

### Instruction Letters

**Purpose** To describe how to perform a task.

**Contents** Instruction letters should include the following:

—The purpose of the instructions

—The instructions, presented sequentially as numbered steps

—Directions regarding what to do if there are problems

**Comments** If the reader does not have a technical background, either do not use technical jargon or provide definitions and explanations as needed. See Figure 3-7.

### Report Letters

**Purposes** To disclose observations and make recommendations, decisions, or conclusions.

**Contents** Report letters should include the following:

—A description of the subject and purpose of the communication

—(If someone other than the reader has asked for the report) The name of the person who requested the report

—Background information and/or observations

—(For non-controversial reports) Decisions, recommendations, or conclusions

—(For controversial reports) *Facts* followed by decisions, recommendations, or conclusions

—A statement indicating your willingness to follow-up, get additional information, hold additional meetings, or other appropriate action

**Comments** The format for a report can be either a memo or a letter (see sections 3-1 and 3-4). In general, a memo format is recommended when the recipient is someone you know and who is already familiar with the subject. See Figure 3-8.

Some pointers for writing reports include the following:

—Keep them short, usually less than five pages

—Keep them simple and to the point

—Remember who your audience is (e.g., do not use jargon and technical terms unless you are sure they will be understood)

## Checking Your Understanding

### Activity D

Write an instruction letter regarding one of the following:

—How to decontaminate a piece of sampling equipment, such as a shovel

—How to program a VCR to tape a one hour program every Thursday evening at 10:00 p.m. on channel 7

### Activity E

Write a letter reporting on the procedure that was required for you to enroll in this class.

# 3-3 General Business Letters

### Concepts

- Environmental technicians may be asked to write business letters that communicate non-technical information, foster good will, address problems, promote business, and request payment that is past due.
- Each of these letters have a set of recommended requirements to make them effective.

## Confirmation, Follow-Up, and Transmittal Letters

These types of business letters are used to communicate information that may or may not be technical.

### Confirmation Letters

**Purpose** To repeat and strengthen the content of a previous communication.

**Contents** Confirmation should include the following:

—A reference(s) to your previous communication(s)

—A summary of the previous communication

—How you can be reached for further information

**Comments**. A confirmation letter can be an effective tool for documenting an oral agreement. An additional step would be to send the reader a copy of your confirmation letter (with a date and signature line at the bottom to be signed) and a pre-addressed, postage-paid envelope. See Figure 3-9.

### Follow-Up Letters

**Purpose** To remind someone to respond to a previous communication.

**Contents** Follow-up letters should include the following:

—A reference(s) to your previous communication(s)

—A summary of the previous communication

—What will happen if the reader does not respond

—How you can be reached for further information

—A thank you for the readers assistance

**Comments** In general, follow-up letters use a more formal or legal tone that is polite, but firm. These letters often reference or include documentation. See Figure 3-10.

### Transmittal Letters

**Purpose** To accompany and introduce other written documents.

**Contents** Transmittal letters should include the following:

—A brief explanation of why you are contacting the company or individual

—A summary of the contents of the letter's enclosure(s)

—How you can be reached for further questions

**Comments** Be sure to highlight a few points from the enclosures that you believe will be of particular interest to the reader. See Figure 3-11.

## Checking Your Understanding

### Activity A

Write a letter of confirmation to your instructor, stating your understanding of the requirements for this course. Include the criteria for making an A, B, and C grade. Be sure to provide a place for the instructor to date and sign the letter.

### Activity B

Several weeks after the Biochemical Company, Inc. was sent the follow-up letter (Figure 3-10) there has been no response. Write a letter informing them of your company's decision to discontinue the use of their products. (Be creative, but polite.)

### Activity C

Prepare a transmittal letter to be submitted with one of your other written assignments in this course. Use the transmittal letter to point your teacher to an area in the assignment that you believe you prepared especially well.

**Land Development Contractors**

21 Shady Tree Road · Rolling Meadows · Vermont 05408 · 802-555-8020

October 1, 1998

George Freeman
The Boring Brothers
335 Mill Creek Road
Nicetown, Vermont 05408

Dear George:

The purpose of this letter is to confirm the details agreed upon during our telephone conversation yesterday. According to the conversation, your company will be available to drill soil borings at the 1415 Jackson Street site in Nicetown on October 15th. You indicated that you would be responsible for arranging to have all underground utilities located and marked in preparation for the October 15th drilling activities. You also indicated that you would provide a certified driller to direct these activities. If you could please provide that person's name and certification number before October 15th it would be greatly appreciated.

Please confirm that these are also your understandings by signing a copy of this letter and returning it to me in the enclosed envelope. The second copy is for your files. Note that a space is provided in the letter for recording the driller's name and certification number. If anything in this letter is different than you understood, please contact me immediately.

Very truly yours,

Dana Byron
Field Technician

Figure 3-9: Confirmation Letter

July 1, 1999

Mr. John James
BioStuff Chemical Company
3501 Industry Road
Akron, Ohio 44311

Dear Mr. James:

On June 1, you were notified, in writing, that two of your company's Material Safety Data Sheets (MSDS) are missing information required by the Occupational Safety and Health Administration's (OSHA) Hazard Communication Standard. I am still seeking to obtain the information on two Biochemical Company, Inc. products. The products and missing MSDS sections are as follows:

1. Biocoat – No health effects section
2. Biocide Solvent – No physical hazards section

Enclosed, please find copies of the two incomplete MSDSs and the letter dated June 1, 1999. To meet our Hazard Communication training requirements, you will need to supply the revised MSDSs by July 14, 1999. If you do not provide the information by that date, I will notify OSHA of the omissions and contact other suppliers for alternative products.

If, for any reason, you cannot comply with this request, please contact me immediately at 1-502-555-2002, X444. Your prompt attention to this matter will be greatly appreciated.

Yours truly,

Al Jackson
Environmental Health & Safety Officer

Enc. (3)

925 Acorn Road, Lone Oak, Kentucky 42002 (502) 555-2002

Figure 3-10: Follow-Up Letter

Fairfield County Water District
765 Deer Run Road, Pittfield, Pennsylvania 15658
(724) 555-4270

July 20, 1998

Virginia Phong
332 Daffodil Drive
Willpen, Pennsylvania 15658

Dear Ms. Phong:

Enclosed please find the recently completed report on the quality of the drinking water from the Fairfield County Water District. We thought that you, as a local school board member, would be particularly interested in the findings that are significant to young children.

On page seven of the report, you will find a table with a complete listing and analysis of the water being distributed to Lincoln and Washington Schools. Of particular interest are the findings that both nitrate and lead levels are well within acceptable drinking water standards for children.

I hope that you find this report reassuring and useful. If you have any questions about the more technical items in the report, please give me a call at the above number.

Sincerely,

Dan Phillips
Safety and Health Coordinator

Enc.

Figure 3-11: Transmittal Letter

## Good Will, Support, Recommendation, and Thank-You Letters

These types of letters are for promoting and maintaining good relations with clients, business associates, and others.

### Good Will Letters

**Purposes** To congratulate, compliment, or express appreciation.

**Contents** Good will letters should include the following:

—Who you are, if the reader does not know you

—What you are congratulating, complimenting, or expressing appreciation for

—The reason(s) why you are complimenting, congratulating, or expressing appreciation

**Comments** Good will letters create and maintain positive relations among business colleagues. They are not only good for human relations, but for business as well. Imagine how each of the opening sentences in Figure 3-12 would make some reader feel.

### Support Letters

**Purposes** To encourage the selection of a proposal or to support a person, idea, or project.

"I wanted to take a moment to tell you how much I appreciate your attention to detail in our professional society meetings. I know that your efforts take time away from your already busy work schedule...."

"I wish you sincere congratulations on your well-deserved recognition as Citizen of the Year...."

"As I was driving by your new office this morning, I was struck by the attractiveness of your renovation efforts and the landscaping. I compliment you on making your dream into reality...."

Figure 3-12: Opening Statements for Good Will Letters

**Contents** Support letters should include the following:

—Who or what you are supporting

—Why you are supporting it or the person

—The benefits (financial, talent, partnership, or other) your support will bring to the proposal or selection

**Comments** If you are writing to a Senator, Congressman, or head of state, the format of your letter may differ from the one described in the opening section of this chapter. For information regarding specific letter writing formats, refer to a style manual that describes how to formally address those in important positions or public office. Figure 3-13 is an example of a letter of support for a proposal.

### Recommendation Letters

**Purpose** To recommend a person for their demonstrated job skills or character.

**Contents** Recommendation letters should include the following:

—Who you are

—How long you have known the person

—In what capacity you have know the person

—The person's strengths

—Why he would be an asset in a particular job

**Comments** Figure 3-14 is an example of a letter of recommendation written on behalf of an applicant for a position.

### Thank-You Letters

**Purpose** To express thanks for something that was given.

**Contents** Thank-you letters should include the following:

—A thank you with a description of the gift or service

—Some concrete detail about how you used or enjoyed whatever was received.

**Comments** The best thank-you letters are warm and appreciative in tone, such as the letter in Figure 3-15.

Partnership for Enviornmental Technology Education
Pleasanton, California 94566

October 14, 1998

Phil Fabian
California Department of Toxic Substance Control
Sacramento, California 95814

Dear Mr. Fabian:

I am writing to express my support for your application in the special volume-production category of the Lead-Based Paint Abatement Grant Program as outlined in the U.S. Department of Housing and Urban Development's Request for Grant Application.

The 600 member Partnership for Environmental Technology Education (PETE) is currently coordinating the vocational training of thousands of students in the community and technical colleges across this country. Since these programs are already providing trained environmental technicians to local industries, consultants, and government agencies, the fit would seem to work well with your proposed lead abatement-action program.

I will support your efforts by facilitating an outreach program through PETE. I am confident that this will enhance our program as well as improve your ability to effectively reach those involved in the proposed lead abatement activities.

I am looking forward to working with you in your efforts to establish a national lead abatement-action program.

Sincerely,

Dick Paulson

Figure 3-13: Support Letter

**XYZ Analytic Lab**
636 Lewis and Clark Highway
Helena, Montana 59604
(406) 555-9604

February 2, 1999

Jackie Powers
Human Resources Department
BioStuff Chemical Company
3501 Industry Road
Maple Valley, Ohio 44311

Dear Ms. Powers:

I am writing this letter on behalf of Jennifer Miller's application for the position of laboratory supervisor. Jennifer has worked for me for the past three years and has been an outstanding laboratory technician. Six months after Jennifer starting work in our lab, her team leadership and writing skills became apparent, and she was promoted to the lead technician position.

Her knowledge of sampling protocol and equipment and her skills in preparing formal laboratory reports are excellent. This knowledge and ability has allowed her to recommend some improvements in our sampling and reporting program that have resulted in considerable savings to our company. Jennifer's work habits are exemplary. She has not missed a single day since she started working here. Her even temperament and pleasant personality makes her a valued employee and an excellent team member.

Although Jennifer will be missed, I would highly recommend her for your position as laboratory supervisor. If I can be of any further assistance, please contact me.

Yours truly,

Henry Remington

Figure 3-14: Recommendation Letter

**Bixler & Bixler**
1200 Main Street
Hanahan, South Carolina, 29410
(843) 555-1200

October 14, 1998

Stan Baker
O'Donnell & Maxwell
55 Hummingbird Lane
Pinehaven, South Carolina 29405

Dear Mr. Baker:

Thank you for your immediate and thorough response to my request for information about Skybart Corner. The maps, written descriptions, and copies of the laboratory reports from your own sampling efforts have been most helpful to my site investigation process. It is always a pleasure to find someone who keeps such organized and detailed records. On behalf of my client, who hopes to receive clearance soon for purchasing the property, I send my warmest thanks for your efforts.

Sincerely,

Todd Bixler

Figure 3-15: Thank You Letter

# Checking Your Understanding

### Activity D

Write a letter of appreciation to the appropriate college official for providing you with an attractive place to go to school. You may select the landscaping, buildings, classroom, or other feature as the topic of your appreciation.

### Activity E

Write a letter of support to the city council for a proposed recycling program. In the letter explain how you and members of your class can be of assistance in starting and expanding the program.

### Activity F

Write a letter of recommendation for a teacher who is applying for a new teaching position.

### Activity G

Write a thank you letter to the reader of the confirmation letter shown in Figure 3-9. Thank the individual for agreeing to get the utilities marked, finding a local certified driller, and preventing confusion on the Jackson Street job.

# Claim, Complaint, Adjustment, and Apology Letters

These types of letters deal with problems. When drafting letters that involve a dispute or could have possible legal ramifications, ask the following questions:

—Do you have the facts straight? Better yet, do you have documentation to back up your claims, such as field notes, telephone log notes, or copies of invoices and receipts?

—Are you irritated at the time of writing? If so, ask someone to read the correspondence carefully, looking for threatening wording or promises you may later regret.

—Is the situation legally threatening? You may need assistance and approval from someone representing your company, such as a manager or an attorney.

## Claim and Complaint Letters

**Purposes** To express dissatisfaction and propose an equitable adjustment or solution.

**Contents** Claim and complaint letters should include the following:

—What has prompted you to write

—The facts associated with the dissatisfaction

—What the situation has cost you or others in time, money, and/or inconvenience.

—What you want the reader to do

—What actions you will take if the matter is not settled

—Your expectations for when the reader will act on the claim or complaint

**Comments** The difference between claim and complaint letters is that complaint letters generally do not ask for economic remuneration, and claim letters do. When writing either of these letters, keep in mind that your choice of wording may interfere with, rather than help, your chances of getting the desired results. Your strategy should be to encourage the reader to take a positive action. For both types of correspondence, be sure that all of the facts are correct.

Note that Figure 3-16 is an example of a short claim letter between two businessmen who are well acquainted asking for money. The letter in Figure 3-17 is an example of a complaint letter making no monetary request.

## Adjustment and Apology Letters

**Purposes** To respond to a claim (adjustment) or to respond to a complaint (apology).

**Contents** Adjustment and apology letters should include the following:

—Reference to the claim or complaint letter

—Whether you agree or disagree with the claim or complaint

—What actions you have taken or are going to take

—An apology or comment of good will

**Comments** There are two types of adjustment letters: one is for claim letters that are correct and the other for those that are not (see Figure 3-18 and Figure 3-19). The goal for both letters is to maintain a good relationship between you and the customer. If you or your company is in error, then thank the customer for

**Big Country Ranch**
Fisherman's Road • Helena Montana 59625 • (406) 555-9773

August 25, 1999

Henry Remington
XYZ Analytic Lab
636 Lewis and Clark Highway
Helena, Montana 59604

Dear Henry:

I just received your statement for the analysis performed on the soil samples from the Wilson property. According to the contract, your laboratory was to charge a flat fee of $75 for each of the thirty samples. On closer inspection, I find the invoice exceeds that amount by $375.

Perhaps your laboratory made an error in reporting the actual number of samples analyzed. As soon as I receive a corrected statement, I will remit payment in full.

Thank you in advance for your prompt attention to this oversight.

Best regards,

Lyle Holt

Figure 3-16: Claim Letter

*Arthur Kramer*
*101 Blue Bird Avenue*
*Maple Valley, Ohio 44311*

May 10, 1999

Mr. Bert Stewart
BioStuff Chemical Company
3501 Industry Road
Maple Valley, Ohio 44311

Dear Mr. Stewart:

Last Tuesday, I observed a group of teenagers from our neighborhood conducting a peaceful protest in front of your plant. They said they were protesting your toxic release inventory (TRI) data, which had been reported in the local newspaper the previous day. In addition to marching near your front gate, they were chanting and carrying signs advocating "Environmental Justice" and "Clear the Air for Everyone."

I believe that the teens were conducting their protest responsibly. As your car approached the plant, I was surprised to see you drive dangerously close to the protesters. I also heard you insult the protesters. I believe you owe these young citizens an apology. If your plant is releasing toxic vapors affecting our health, then you must stop it immediately. If you do not apologize, I will encourage the protesters to bring this matter to the attention of our city and state government officials.

Sincerely,

Arthur Kramer

Figure 3-17: Complaint Letter

**XYZ Analytic Lab**
636 Lewis and Clark Highway
Helena, Montana 59604
(406) 555-9604

August 30, 1999

Lyle Holt
Big Country Ranch
Fisherman's Road
Helena, Montana 59625

Dear Lyle:

This letter is to acknowledge that your statement of August 15, 1999 did include a $375 overcharge. Thank you for calling it to my attention. One of our new billing department employees failed to realize that the current fee structure does not apply to our long-valued customers.

A statement showing the corrected amount will be sent. We appreciate your business and hope that we continue to be your choice for quality analytical services.

Yours truly,

Henry Remington

Figure 3-18: Adjustment Letter in Agreement with a Customer

**XYZ Analytic Lab**
636 Lewis and Clark Highway
Helena, Montana 59604
(406) 555-9604

August 30, 1999

Lyle Holt
Big Country Ranch
Fisherman's Road
Helena, Montana 59625

Dear Lyle:

It is always difficult when a breakdown in communication within your company happens. After consulting our records, I found that the technician who delivered your samples from the Wilson property requested that Sample 36S-4W-99 be analyzed for PCBs and reported to ppb accuracy. That is why your statement of August 15, 1999 shows the additional $375 charge.

Since this request was beyond your contract and our instrumental capabilities, the laboratory manager sent the sample out for the requested analysis. Because you are a valued customer, we just passed their charge of $375 for the analysis on to you. I trust this satisfactorily explains the additional charge in question.

As always, it has been a pleasure serving you and we hope to continue providing you with our quality analytical services in the future.

Yours truly,

Henry Remington

Figure 3-19: Adjustment Letter in Disagreement with a Customer

**Evergreen Country Club**
Pine Hills Road
Bozeman, Montana 59717
406•555•9717

August 9, 1999

XYZ Analytic Lab
636 Lewis and Clark Highway
Helena, Montana 59604

The building committee of Evergreen Country Club would like to express its sincere apologies for the delay in responding to your correspondence of 6/26/99 and subsequent letters dated 7/15/99 and 8/2/99. As you know, we are also in the process of making improvements to the Clubhouse and as a result our office has been in a state of disarray. Apparently, your request for specifications and follow-up letters were put in a moving box and shipped to another office.

Please accept our simple explanation for this most embarrassing oversight. We were very happy with your services and the professional manner in which you have handled this matter. Enclosed please find the specifications that you requested. We are looking forward to receiving your report and apologize for any inconvenience that this may have caused.

Sincerely,

Betty Rockford
Chairperson

Figure 3-20: Apology Letter

bringing it to your attention, state the adjustment or correction that will be made, and apologize for any inconvenience that it may have caused. You might even consider providing a brief explanation of the circumstances that resulted in the error.

If you or your company is not in error, and you must disagree with the customer's request for an adjustment, then tactfully present an explanation for the possible misunderstanding. Always close the letter graciously with a positive statement, such as "It is always a pleasure to serve you."

In many ways, a letter of apology (see Figure 3-20) is similar to a letter of adjustment in which you are accepting responsibility for an error. The difference is no monetary remuneration is offered. Your purpose for writing such a letter should be to express regret for the error and to retain a business relationship with the reader.

## Checking Your Understanding

**Scenario**

Last month you purchased two Toothy® blades for your 48" Sawz Things® band saw. One blade had 14- and the other 18-teeth per inch, and both were recommended for cutting soft metals, like aluminum, and plastics. After carefully following the directions and installing the 14-tooth blade, you proceeded to make your first cut. After sawing only about two inches, the blade began to smoke, changed color, and got hot. By the time you sawed another inch, the blade snapped, resulting in some damage to the plastic you were cutting, but none to your fingers.

Undaunted, you proceeded to install the other blade and started the cut again. This blade also started to smoke and discolor, so you turned the saw off. After stopping the saw, you removed the blade and went back to the hardware store where you had purchased the blades for $15.90 each. The store manager not only refused to replace the blades with another brand, but also refused to refund your money or give you a store credit. He said your complaint was with the manufacturer and that you should return the damaged blades to them along with a letter asking them to respond to the complaint.

### Activity H

You are angry with both the store manager and the manufacturer in the scenario. Write a claim letter to the manufacturer, with a copy to the store manager, and explain why you should be fully reimbursed for the blades.

### Activity I

Pretend to be the store manager and write a letter of adjustment for the incident in the scenario, explaining why the store took the position it did, while attempting to keep the customer.

**Scenario**

You are the owner of a small Minnesota-based company that routinely provides soil-sampling services. You are working as a contractor for a large Florida developer who wants to construct a commercial building on a site where a full-service gasoline station was located. Some visible soil discoloration suggests that an underground tank may have leaked in the past. The local building department requires that soil sampling and analysis be completed before the developer can start construction.

This developer has recently built on other properties in the area. He is eager to get started on this project. You have signed a contract stating that your sampling activities will be completed by March 1, 1999. An unusually cold February has frozen the ground solid for several feet down. On two occasions, you attempted unsuccessfully to obtain the soil samples. You have a signed contract for $5,000 and are afraid that you will not be able to meet your contractual obligations on schedule. You would like to retain the developer as a customer.

### Activity J

It is now the beginning of April and no soil samples have been taken. Pretend you are the developer in the scenario and write a letter of complaint to the soil-sampling service.

### Activity K

Write the developer in the scenario an apology letter explaining why there has been a delay. Also, explain why it is in his best interest to use your services in the future.

### Activity L

The laboratory that analyzes your soil samples just sent a letter informing you that because of business volume you have become one of their "preferred customers." Along with this status, you will receive an additional 10 percent discount on all analysis charges. Write the developer an adjustment letter explaining the delay and offer to make an adjustment in the contracted price.

## Sales and Collection Letters

Typically, these types of letters are used by environmental technicians operating their own businesses.

### Sales Letters

**Purposes** To sell services and increase customers.

**Contents** Sales letters should include the following:

—An attention getting statement that motivates the reader to finish the remainder of the letter

—How the reader can benefit from your services and/or why the reader should use your services

—A statement telling the reader what immediate action to take, such as call a toll-free number or mail back a card

**Comments** The hook or first sentence of a sales letter that grabs the reader's interest can take on many forms, such as stating a problem (e.g., "Are you spending more money than you should disposing of waste oil?") or the presentation of an interesting fact (e.g., "Five recycled PET bottles make enough fiberfill to stuff a ski jacket.").

Use persuasive language throughout the letter, and emphasize the advantages of using you and your company's services.

Keep the letter short, providing only enough information to get the reader to contact you. Pre-addressed, postage paid mail-backs, toll-free telephone numbers, and web sites are ways to make it easier for readers to request additional information. Refer to Figure 3-2 for an example of an effective sales letter.

### Collection Letters

**Purpose** To collect money for services rendered while maintaining good relations with a client.

**Contents** Collection letters should include the following:

—A brief reference to the date of the bill, services rendered, and the amount due

—(First letter) A friendly reminder that the bill is past due

—(Second and third letters) A reference to the number of times a collection letter has been sent

—(Second and third letters) A summary of the contract or agreement

—(Second and third letters) A description of the services you have rendered

—(Third letter) What actions you will take if payment is not made by a certain date

—A thank you for the reader's prompt attention to the matter

**Comments** The tone of a first collection letter should be that of a friendly reminder giving the customer the benefit of the doubt. An example:

> "Your check may have crossed this letter in the mail. If so, please disregard this reminder and know that we appreciate your business."

The tone of the second letter should be more firm. An example:

> "This is a second reminder that your payment for the site investigation is overdue. As agreed in our contract, we have paid the drilling and laboratory subcontractors. We have met all of the terms in our agreement and now ask that you fulfill your obligation by sending us the $6,000 payment."

The tone of the third letter should become more formal and legal. State what action you are prepared to take, and give a time limit for payment. An example:

> "Your account is now 90 days overdue, and we have not received any communication from you about this bill. If we have not received your payment by June 15th, then we will turn this matter over to our attorney."

Whenever a letter concerns a challenging situation or promises to be complex in its details, outlining may help organize your thoughts. For example, suppose that a client is 60 days overdue on a bill. The sequence of events is the following:

—The customer is the building committee for the local country club.

—They hired you to conduct a Phase I assessment on agricultural property that they plan to use for an 18-hole addition to the their golf course.

ECS ECS ECS ECS

**ENVIRONMENTAL CONSULTING SERVICE**

100 Sockeye Road ■ Salmon, Idaho 83467 ■ 208-555-0001 ■ fwilliams@ecs.com

August 1, 1999

Betty Rockford, Chairperson
Evergreen Country Club
Piney Hills Road
Bozeman, Montana 59717

Dear Ms. Rockford:

It is possible a misunderstanding has occurred regarding the committee's obligation to pay for my Phase I consulting services. Your payment for these services is now sixty days overdue. The initial invoice was sent on May 3rd, a follow-up invoice on June 1st, and a letter reminding you of your obligation on July 5, 1999.

The May 3rd invoice was sent following a conference call in which we agreed I would conclude my role in the Phase I assessment. It was my understanding your committee had decided that it could save money by having some of the members continue the investigation. In that conversation, I told you that I had already spent some time on the assessment, and I recall that you said I was to send you a bill. You also said that if your Phase I assessment indicated any potential contamination at the site, you may rehire me for a Phase II site investigation.

At the time you took over the Phase I research, I had invested a total of eight hours. The agreed-upon rate was $60 per hour, for a total of $480. This time was spent accomplishing the following tasks:

— Obtaining documents
— Interviewing people
— Studying and organizing data
— Meeting with the committee

You also asked for any documents I may have obtained from my efforts, to save you time in conducting your own research. I would be happy to provide those upon receipt of payment.

I am sure that you are all people who are fair and prompt in paying your bills. There is probably some simple explanation for this oversight, and I would greatly appreciate you putting the check in the mail today.

If your Phase I research does indicate that you need a Phase II site investigation, I would be pleased to work with your committee again.

Sincerely,

Fred Williams
Field Technician

Figure 3-21: Collection Letter

—They decided, after you had begun the Phase I research, that various committee members could take over the assessment and save money.

—The committee chairman stated in a telephone conversation that the committee would pay you for the time you had already put into the research.

—The committee has ignored 1) the original invoice for your hours, 2) a reminder invoice, and 3) a brief, friendly letter asking for the payment to be made.

—The committee chairman said they might rehire you for a Phase II site investigation, if the Phase I results show that one is needed.

—The committee asked you for all the documents that you obtained during the hours you spent on the project.

To organize your thoughts, jot down the following points and then arrange them so the letter will flow smoothly from idea to idea.

| List of Points to Cover | Organizational Sequence |
|---|---|
| Repeat the way you understood the phone call | 2 |
| Give them the benefit of the doubt–maybe they've been busy | 1 |
| List the tasks you have already done and the hours spent | 4 |
| Compare to how they would feel in this situation | 5 |
| Keep open the chance to do a Phase II, if they need it | 6 |
| List the attempts at collecting the payment | 3 |

Next, research your daily log of activities to verify that your billing hours are correct.

The log shows:

—One hour contacting the United States Geological Survey (USGS) for a map and the United States Department of Agriculture Farm Service Administration (USDA FSA) and the State Department of Transportation (DOT) for current and historical aerial photos.

—Two hours in telephone interviews with the owner and two neighbors.

—One hour studying the map and photos after they arrived.

—Four hours in telephone conversations and meetings with the committee.

The last step is to write a letter that will help you to receive the overdue payment and keep the committee as a future client. See Figure 3-21.

## Checking Your Understanding

### Activity M

Select one service or skill you would be able to provide a customer as a result of completing this class. Pretend that you have started your own business to supply this service/skill. Write a sales letter.

### Activity N

Two months after having written Figure 3-16, and having received the response in Figure 3-19, you still have not received payment for the extra $375. Now the lab needs to write a letter of collection to the contractor.

# 3-4 Memos

## Concept

- Business memos must follow a format that includes a date, the name of the individuals sending the receiving the memo, a subject line that encapsulates what the memo is about, a body of text, and a list of the names of any attachments
- The three main purposes of writing a memo are to inform, to persuade, and to document events.

## Memo Format

Memoranda, referred to as memos, are letters sent to colleagues within the same company. There are many acceptable styles for these communications. In fact, businesses often adopt one style for company-wide use. In general, business memos consist of three parts:

—**Heading** The memo heading contains the following lines:

- **To:/From:** For formal memos, use formal titles next to the names, even if you are on a first name basis with the readers. Titles are not necessary for informal memos. Always be sure that you have addressed the readers by their correct names, checked spellings, and taken the time to determine the correct job titles.
- **Date :** The date the memo is written.
- **Copy/Blind Copy:** (May also be written as cc:/ bc:.) Copy is used if you want to send copies of the memo to other interested persons and to make the persons to whom the memo is addressed aware that others are reading the memo. Blind copies are copies sent to additional people without the recipient's knowledge.
- **Subject:** The subject line highlights what the memo is about, as discussed in Section 3-1.

—**Body** The body, which is the memo's content, begins two lines below the heading.

—**Attachments (Optional)** If there are attachments, then they should be listed, by title, at the end of the memo so the reader is made aware of any accompanying information and can follow up if any attachments are missing. Attachments are additional documents that provide further, more detailed supporting information. They can also consist of graphs, lists of information, tables, etc. For documentation memos (see description letter in this section), attachments are particularly important because they become part of the preserved record of events and link all supporting materials to one document.

In 1985, I became department head of Industrial Technology at a small community college. It was a floundering conglomeration of technical programs ranging from automotive technology to cosmetology. To get a handle on some of the problems facing the department, I turned to the employers of our graduates.

Interestingly enough, the main complaint these employers had was not the technical skills of graduates but their communication skills. Employers wailed about graduates who could not compose a complete sentence, who could not spell, and who could not organize or express their ideas clearly. In one instance, an employer complained that two of our graduates had great technical skills, but he could not promote them to management positions because of their poor writing skills.

If your writing skills could use some polish, I encourage you to endure a few extra writing courses before you graduate. Colleges offer traditional writing classes as well as ones that focus on technical or business writing. If your schedule can't accommodate any more time in a traditional classroom, there are hosts of very good, self-paced writing courses currently available through the Internet. (To locate these classes, search via "Business Writing Courses" or "Technical Writing Courses.")

Finally, don't think of the writing courses you are required to take for graduation as torture sessions enjoyed only by the English faculty who teach them. Rather, think of writing courses as high return career investments that have the potential of helping you stand out in the field of applicants or provide that extra edge for promotion.

Lea Campbell, Regional Director
*South Central Partnership for Environmental Technology Education*

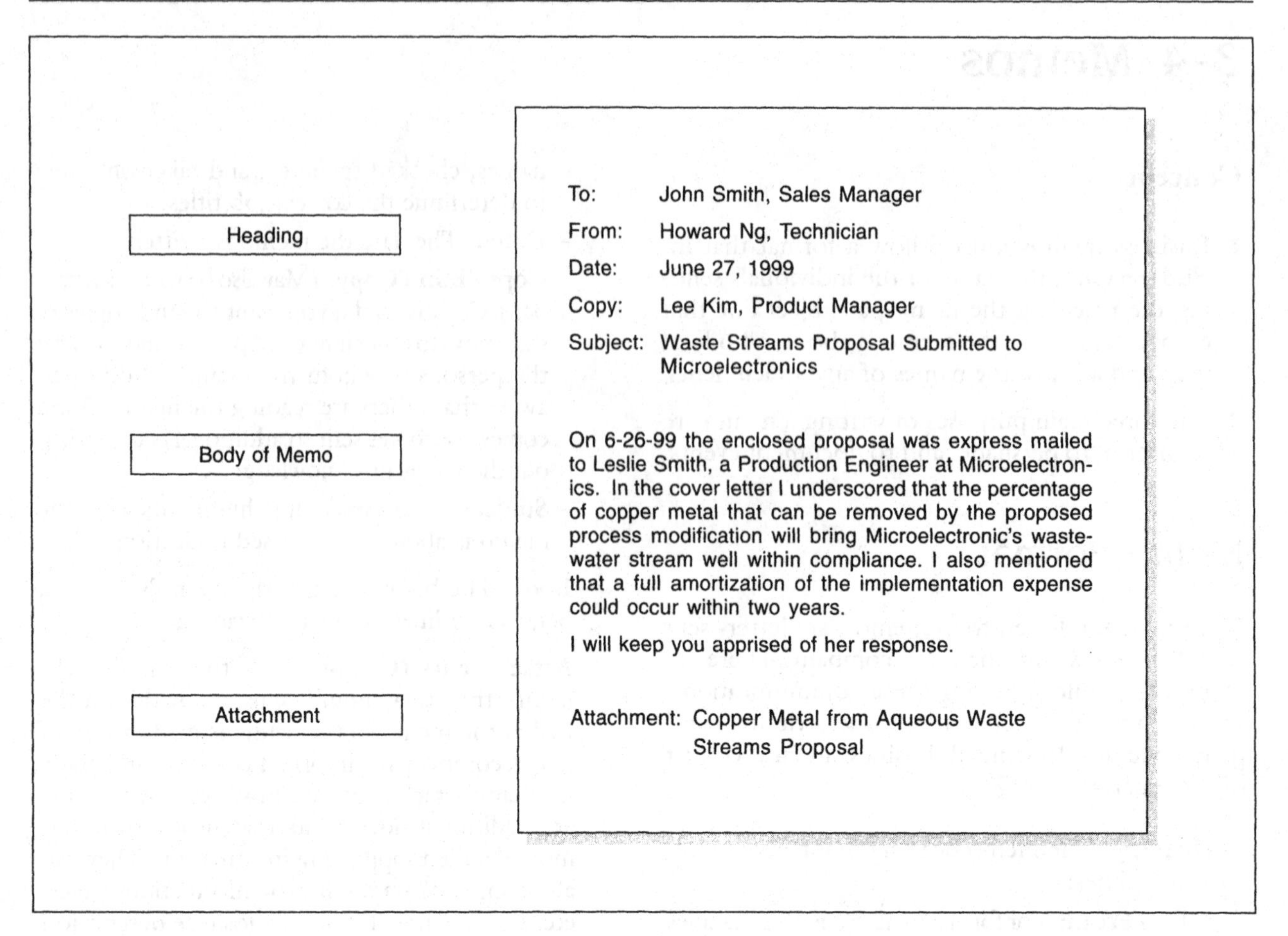


To: John Smith, Sales Manager
From: Howard Ng, Technician
Date: June 27, 1999
Copy: Lee Kim, Product Manager
Subject: Waste Streams Proposal Submitted to Microelectronics

On 6-26-99 the enclosed proposal was express mailed to Leslie Smith, a Production Engineer at Microelectronics. In the cover letter I underscored that the percentage of copper metal that can be removed by the proposed process modification will bring Microelectronic's wastewater stream well within compliance. I also mentioned that a full amortization of the implementation expense could occur within two years.

I will keep you apprised of her response.

Attachment: Copper Metal from Aqueous Waste Streams Proposal

Figure 3-22: Memo Format

## Body of the Memo

In general, a memo consists of an opening statement and a discussion. The opening statement usually communicates the purpose of the memo, some brief background information, and the specific course of action the reader is being asked to take. It should be clear and concise and provide the reader with only the needed information. For example, a memo with a subject line of "1998 Annual Safety Training Schedule" might have the following opening statement:

> "As you know, all employees are required to attend one safety refresher course each year. The 1998 Safety Training Schedule is listed below. Please review the schedule with the employees in your department and have them each select a session to attend. Provide a list of employees and their session dates to the Health and Safety Division no later than Friday, January 15."

The discussion segment of a memo is where all of the detailed information is communicated. A discussion has one of the following purposes:

—To inform

—To persuade

—To document a certain course of action

**In an informative memo** (see Figure 3-23), which is used to acquaint the reader with business or technical information, the discussion segment is usually very direct, starting with the most important points and then moving through to the supporting details.

**A persuasive memo** (see Figure 3-24), where the reader is persuaded to adopt a certain course of action, is usually more complex. The discussion segment is lengthy and may be less direct. This type of memo often begins with a background statement,

**NST Never-Spill Transport Inc.**

**MEMORANDUM**

**To:** Fred Hamilton, Supervisor, Transport Services Division

**From:** Emily Ritter, Asst. Supervisor, Environmental Division

**Date:** June 17, 1999

**Subject:** Revision of Product Onloading and Offloading Policy

This memo is to make you aware of a revision in company policy as it pertains to the loading and unloading of product. Please review this policy with the employees in your division and have them complete the attached forms acknowledging that they have received, read, and understood the new policy. The forms should be completed by June 23, 1999, and retained in your office.

As you are aware, a transport truck was recently involved in a major spill at loading dock three and an undetermined amount of raw product was washed via a storm drain into the Little Squaw River. A number of factors contributed to that spill, including poorly maintained hoses, failure to cover a storm drain grate, and the fact that the driver of the truck was not present while product was being unloaded. To prevent future incidents of this nature, the following policy is effective immediately:

1. All transport trucks will now carry storm drain cover mats and spill booms. Prior to pumping, all drains in the vicinity should be covered and have a spill containment boom placed securely around them.

2. Pumps, valves, and hoses on transport trucks will be inspected for wear to ensure they are in proper working order before employees begin pumping product. Pumping should not begin until all equipment is functioning properly and worn equipment or hoses are replaced.

3. Emergency shut-off valves will be installed on all transport trucks and will be inspected prior to pumping to ensure that they are in proper working order.

4. The driver must remain with the transport truck at all times while product is being pumped on or off.

As in the past, if a spill occurs, employees should take immediate action to close emergency shut-off valves and turn off pumps. The spill should be contained as much as safely possible, and Environmental Services should be paged at X4221.

If you have questions or need additional assistance implementing this policy, please contact me at X4222 or J. P. Brian at X4223.

Attachments: Form 2361-Acknowledgment of Policy Training (23 copies)

Figure 3-23: Informative Memo Example

## MEMO

To: Paul Aragon, Training Manager
From: Sara Woo, Environmental/Safety Trainer
Date: February 15, 1999
Subject: 40-hr. HAZWOPER Contract Training Bids

In January, you asked me to research local contract trainers and solicit bids from trainers with the ability to take over all 40-hour HAZWOPER training for new employees at the Jameson, Louisiana, plant. I have located three contractors able to meet our training needs and have attached their bids.

Below is a summary of my research findings along with a recommendation that we contract with Carrie Hills Community College for this training.

— Poole & Associates, Inc.
Bid Amount: $750 per person with a 10% discount on any class with over 10 trainees

Poole & Associates is the oldest and most established of the contract trainers in the area. James Poole, the owner, has over 15 years in contract training and while he has provided excellent training to this company in the past, Poole turned in the high bid.

— Martin Environmental Training, Inc.
Bid Amount: $695 per person with a 10% discount on any class with over 10 trainees

Martin Environmental Training, Inc. is owned and operated by Terry Martin, formally of Helena Petroleum. Martin Environmental is new to the training industry, and while I found Mr. Martin to be both enthusiastic and knowledgeable, the company had a limited amount of training equipment. I feel that Mr. Martin would be able to provide quality lecture training but the limited quantity of equipment would greatly restrict the amount of hands-on training new employees would receive.

— Carrie Hills Community College, Hazardous Materials Management Program (non-credit)
Bid Amount: $600 per person

Carrie Hills Community College has been doing contract training for local industry for the past 5 years. I checked with a number of plants that are currently contracting with Carrie Hills, and, without exception, the training managers were extremely pleased with the level of training their employees received. The training staff members are all National Environmental Training Association Certified Environmental Trainers, and the program offers access to a large collection of equipment, including computer-based safety training. An additional advantage of working with the Carrie Hills program is the access employees would have to the degree program in Hazardous Materials Management. Carrie Hills is able to assist employees wishing to pursue a certificate in Hazardous Materials Management by offering evening classes and classes on site if a group of 10 or more students can be identified.

At this time, I would like to recommend that we extend a 6-month contract to Carrie Hills based on the following:

— The reputation this institution has earned within local industry
— The level of certification held by Carrie Hills faculty
— Carrie Hills' low bid
— The additional education opportunities Carrie Hills can extend to plant employees.

If you have any additional questions, please let me know.

Attachments: Training Proposals from
Poole and Associates, Inc.
Martin Environmental Training, Inc.
Carrie Hills Community College

Figure 3-24: Persuasion Memo Example

**Petro Products**

## MEMORANDUM

To: Joe Green, Plant Manager
From: Jay Chang, Head of Security
Date: September 14, 1999
CC: Jeff Gray, Environmental/Safety Manager
Subject: Speed Limit Violation: Dole Contracting

This memo is to document recent speed limit violations by employees of Dole Contracting while in the Port Boyuea facility. On 12/3/98, Don Jones, security officer, issued warning statements to two Dole contracting employees. In both instances, Dole employees were operating private vehicles within the plant at speeds well in excess of the posted 15-mph limit. At the time, Dole management was warned that if additional instances of speeding by Dole employees occurred, Dole employees would be barred from bringing their vehicles into the plant.

On 1/23/99, a Dole employee lost control of a Dole company vehicle and crashed into a drainage ditch. The driver received minor injuries. An investigation of the crash revealed that he was traveling at a minimum speed of 35 mph. The posted speed on the stretch of plant-maintained road where the accident occurred was 10 mph.

On 2/1/99 Dole management was advised of the results of the accident investigation, apprised of plant policy in regard to speed violations, and notified that Dole employees would no longer be allowed to operate vehicles within the plant.

At this time Charles Coleman, president of Dole Contracting is appealing the barring of Dole vehicles from the plant. He contends that not allowing Dole vehicles into the facility will seriously hinder his company's ability to honor current contracts. He would like to request a meeting with you at your earliest convenience.

Please let me know if and when you would like to meet with Mr. Coleman.

Attached you will find copies of supporting documents.

| | |
|---|---|
| 12/3/98 | Speed Violation Reports |
| 12/5/98 | Memo to C. Coleman |
| 1/3/99 | Accident Investigation Report |
| 2/1/99 | Memo to C. Coleman |
| 2/5/99 | Dole Letter of Appeal |

Figure 3-25: Documentation Memo Example

which provides a history of the problem the business is attempting to solve and the actions and results thus far achieved. Normally, the background section is written chronologically, describing what happened first with subsequent events recounted in the order in which they occurred. Note that in persuasive memos you should never overstate the case or predict dire consequences unless those consequences are a real possibility.

Following the background is a statement that outlines the course of action you are recommending and the rationale for it. In this section, you might want to use a format that begins with the strongest and most important justification, followed by the weaker and less important reasons.

Sometimes you may want to explain why you recommend one course of action over several alternative plans. In this case, state the recommended course of action, then list the other alternatives and why you discounted them as solutions. Listing alternatives makes reading easier because the beginning and end of each alternative are clearly marked and important information can be found quickly.

**Documentation memos** (see Figure 3-25) are used to record a course of action that was taken. This type of memo is for situations where potential legal action may be taken against the company, or for when disciplinary action has been taken against an employee or division. Remember that even though it may be several months or even years before legal or disciplinary action occurs, documentation is critical because it preserves key pieces of information against the passage of time.

Suppose, for example, a safety manager dismisses an employee for failure to follow company policy. The manager is concerned that the dismissed worker might file for unemployment benefits or charge that the company engaged in some form of discrimination. In such a case, the manager would be wise to write a memo to document the company's efforts to ensure that the dismissed worker was aware of company policy. He would also document the names of the individuals involved and the exact events that led to the dismissal of the worker.

Since a documentation memo is a formal record of facts, the memo for the situation described above would include the full names and titles of the individuals involved, a chronological narrative of actions, and, where appropriate, a record of words exchanged between employees. Opinions and speculations would be excluded.

**For all three types of memos**, the closing statement should be a courteous ending that reinforces the purpose of the memo. A closing statement might include an offer to help complete the subject task or a list of sources the reader can turn to for additional information.

### Before the Memo Goes Out

Regardless of whether the memo is intended to inform, persuade, or document, the basic rules of good writing must still be enforced. Even when the message is obviously important or the ideas are useful, readers tend to discount a memo that is sloppy, contains poor grammar, or has misspelled words. Always edit and proofread (see Chapter 2) memos before they are sent out. Above all, make sure the information in the memo is correct.

## Checking Your Understanding

### Activity A

Read the following scenario and write an information memo that Russ Marquez might send to the EDS employees.

#### Scenario

Russ Marquez needs to send a memo to all the employees in the Engineering and Design Systems (EDS) department. He must remind them of a company policy that requires employees to attend an annual eight-hour HAZWOPER update. Training sessions are scheduled for 8 a.m. to 5 p.m. on January 7, January 14, January 21, and January 28. Russ wants the employees to select the date they plan to attend the training and report back to him no later than December 15. The training will be held in Room 512 in Building 5, and lunch will be provided.

### Activity B

Read the following scenario and write a documentation memo that Sara McCord might send to the Director of Human Resources, Manny Johnson. Don't forget to include any attachments that should accompany the memo.

### Scenario

For two years, James Pinkerton has been working for a large oil company as a hole-watcher on a boiler refurbishment job. A hole watcher is someone who warns workers of danger when they have entered a confined space, and notifies rescue teams should those workers become injured or trapped. As part of his new employee training, James had participated in two weeks of safety training, which included a discussion of actions that are grounds for immediate dismissal. (Sleeping on the job is one of those offenses.) At the conclusion of the training, James signed a statement acknowledging that he received a copy of company policies, participated in the training, and had read the policies and understood the training. In addition, he has a current 40-hour certificate in Hazardous Waste Operations and Emergency Response (HAZWOPER) and participated in the latest annual company safety/ environmental refresher course.

On February 14, a senior engineer, Paul Sandoval, entered the boiler to which James was assigned. Paul claims that when he entered the boiler to inspect work done by other contractors, James was awake and aware that the engineer was entering the boiler. However, when Paul climbed out of the boiler 30 minutes later, the hole-watcher was sound asleep. Paul knew James was truly asleep because the hole-watcher's hard hat was pulled down over his eyes and he was snoring loudly. An engineer's assistant, Wiley Snyder, was also present, and witnessed the incident.

Paul reported the incident to Sara McCord, the manager of the oil company's Safety and Environmental Division. Upon checking the company records, Sara was assured that James knew of the danger that sleeping on the job presented to other workers. She was also sure that he was well aware of the consequences of being caught sleeping. Sara then summoned James to her office. She asked James if he was sleeping on the job. When he admitted he was, she explained that she had no choice but to fire him. Security escorted him from the plant. On February 15, Sara sent a certified letter to James formally notifying him that he was dismissed for sleeping on the job.

## Activity C

Read the following scenario and write a persuasive memo that Pete might send to Ms. Carlos.

### Scenario

Pete Greere was asked by his division supervisor, Rosa Carlos, to research and recommend a new software package for tracking shipping manifests. Pete read literature on 15 different packages and then selected three for actual evaluation. *Waste Trace* was a new manifest tracking system that cost $225 per station. Using a trial edition, Pete found this system extremely difficult to use, and technical support was impossible to reach. *Manifests-Made-Easy* did live up to its name; it was extremely easy to use. However, Pete found that it crashed often and appeared to have a number of annoying software bugs. The cost per station was $400. Pete felt that the third package, *Waste Tracker,* best met the needs of the department. It provided a complete manifest tracking system that was fairly easy to use, came with 40 hours of free technical support and training, and was compatible with the plant's word processing software. The cost of the system was $375 per station.

## Activity D

Many of the word processing software packages available today have a template feature, which automatically formats your memo into a professional looking document. To discover how these features work, you need access to a computer. If you do not have a personal computer, go to the computer lab at your school or local public library. Click the help button on the word processing program and type in either "memo" or "template." After learning how the program performs, list the advantages and disadvantages of using these templates.

## Activity E

Read Scenario 1 in the Appendix 1.

Mr. Crabgrass asked you to send him an information memo (copy to Arnold Turf) summarizing the violations you observed during your first day at Locust®. Use bullets to organize the information about the potential violations.

# Summary

On the job, environmental technicians are expected to write letters for a variety of reasons, such as to explain, report, confirm, instruct, request, compliment, complain, and apologize. Regardless of the reason for the letter, always begin by analyzing its purpose and audience. If the letter is long or complicated, use an outline to organize your thoughts.

Memos are essentially letters to colleagues within a company. Memos are used to inform, persuade, and document. As with letters, they should be error free, polite, factual, and concise.

# 4

# Technical Documents

## Chapter Objectives

Upon completing this chapter, the student will be able to:

1. **Demonstrate** technical skills for writing contingency plans, observational reports, standard operating procedures (SOP), and instructions.
2. **Explain** the differences between standard operating procedures and instructions.
3. **Demonstrate** technical writing skills for recordkeeping in logbooks and field notes.
4. **Explain** how different types of written communications may serve to document work activities and decisions.

## Chapter Sections

**4–1** Types of Technical Documents

**4–2** Contingency Plans

**4–3** Observation Reports

**4–4** SOPs and Instructions

**4–5** Logbooks and Field Notes

# 4-1 Types of Technical Documents

## Concepts

■ Depending on the company or agency for which they work, environmental technicians may be expected to write contingency plans, observation reports, SOPs, logbook entries, and field notes.

The technical documents that are written by environmental technicians are as follows:

—Contingency plans

—Observation reports

—SOPs and instructions

—Logbook entries and field notes

Note that not every company expects environmental technicians to write all of the above documents. As a matter of fact, some environmental technician positions do not require many writing responsibilities. For example, you may work largely in the field with an engineer, scientist, or manager who handles all of the writing, including the field notebook entries. In other jobs you may share many of the writing responsibilities with colleagues or be asked to write all of the technical documents. For example, technicians in small companies who handle all of the environmental health and safety activities usually write the related technical documents. Regardless of the number of technical documents required in a job, writing skills are of great value to employers because each one must be written correctly.

Following are examples of the six types of technical documents:

—**Contingency Plans** A contingency plan regarding managers' responsibilities in the event of a hazardous materials accident.

—**Observation Reports** An Exception Report for the EPA that explains efforts on behalf of a client to track down an apparently lost shipment of hazardous waste. A report regarding the effectiveness of a safety procedure used in the transport of cleaning solvents.

—**SOPs and Instructions** A SOP and instructions regarding respirators that include OSHA's latest respirator requirements in the Code of Federal Regulations (CFR). The SOP outlines the client's policies regarding the use of respirators. The instructions illustrate how to wear and maintain particular models of respirators in use at the company.

—**Logbooks and Field Notes** Log of monthly contaminant level readings around workstation areas. Notes taken in the field describing the monitoring instruments used, the workstations monitored, and the time spent.

## Checking Your Understanding

### Activity A

If you have access to an environmental professional, ask them to tell you about the various writing responsibilities that are a part of their job. Write a brief description of your findings and be prepared to share it with your classmates. If possible, obtain copies of several different types of documents to share with the class.

### Activity B

Pretend that you are an experienced environmental technician mentoring an environmental technology student who asks you to provide a summary of a typical day at your job. Use the information from Scenario 2 in Appendix 1 to draft a response.

# 4-2 Contingency Plans

## Concepts

- Contingency plans are documented preparation plans required by law for responding to accidents and emergencies in the work place.
- Contingency plans must be based on Code of Federal Regulations (CFR) requirements.
- Although contingency plans vary from company to company, most plans must include the EPA's SARA Title III Planning Requirements and OSHA's HAZWOPER Emergency Response Plan Elements.
- Contingency plans should be kept in a binder and be updated on a regular basis.

Contingency plans describe the preparations federal agencies require by law for various accidents and emergencies in the work place (e.g., a hazardous material spill or a fire). These contingency plans, sometimes referred to as written programs, are found in the Code of Federal Regulations (see Figure 4-1).

| Agency | Contingency Plan / CFR |
| --- | --- |
| EPA | Federal, state, and local emergency planning<br>40 CFR 300.215 |
| | Hazardous waste management regulations<br>40 CFR Part 264 Subpart D and Part 265 Subpart D |
| | Risk management programs for chemical accidental release prevention<br>40 CFR Part 68 |
| | Spill prevention, control, and counter-measures (SPCC) planning<br>40 CFR 112 |
| OSHA | Employee emergency plans and fire prevention plans<br>29 CFR 1910.38(a) |
| | Process safety management of highly hazardous chemicals<br>29 CFR 1910.119(n) |
| EPA and OSHA jointly | Hazardous waste operations and emergency response<br>29 CFR 1910.120(q) |

Figure 4-1: Various Contingency Plans Required by the CFR

For more information about the CFR, refer to Chapter 2.

The required contingency plans vary from company to company. Depending on the materials used and the manufacturing process, these plans can consist of a minimum set of the basic contingency plans that are requisite for most companies or a basic plan plus twenty or more other contingency plans. For example, OSHA's General Industry Standards could require contingency plans with such titles as Process Safety Management, Permit-Required Confined Space Entry, Lockout-Tagout, Air Contaminants, Bloodborne Pathogens, Hazard Communication, Means of Egress and Emergency Procedures, Chemical Hygiene, and Hazardous Waste and Emergency Response.

An environmental technician could be expected to write a company's contingency plans from scratch or add to and update existing ones. In general, the best way to create a new set of contingency plans is to start with the basic contingency plans required by the federal government for nearly all businesses. These are the contingency plan requirements found in OSHA's HAZWOPER (Hazardous Waste Operations and Emergency Response) standard and in the EPA's SARA Title III emergency planning regulations. Figure 4-2 and Figure 4-3 list the planning topics required by each regulation.

After a basic contingency plan has been written, other contingency plans may be added after it, such as the Employee Emergency Plan and Fire Prevention Plan required in 29 CFR 1910.38(a). As more contingency plans are written, there may be duplications of information. If this happens, either repeat the information or cross-reference it.

Although there is no generally recognized standard format for contingency plans, an outline format is recommended as shown in Figure 4-4, which contains the first few pages of a contingency plan used by Monsanto Agricultural Company in Muscatine, Iowa. The Muscatine plant produces agricultural herbicides, some of which contain highly hazardous materials, and has an emergency response team, an on-site control center, and a mobile communications center. Note that for brevity, maps are not included for sections 6-5 through 6-9. Normally they would appear as part of the plan.

Each set of plans should be put in a binder or set of binders separated by tabs. (see Figure 4-6). Law requires that contingency plans be updated as regulations change.

| Planning Topics Required by EPA | |
|---|---|
| (1) | Transportation routes; additional facilities (e.g., hospitals) at risk due to proximity |
| (2) | Response methods and procedures |
| (3) | Designation of community and facility emergency coordinators |
| (4) | Notification procedures in event of a hazardous materials release |
| (5) | Methods of determining occurrence of a release and the area or population likely to be affected |
| (6) | A description of emergency equipment and facilities and persons responsible for them |
| (7) | Evacuation plans and alternative traffic routes |
| (8) | Training programs and schedules |
| (9) | Methods and schedules for emergency response exercises |

Figure 4-2: Excerpt from EPA's SARA Title III Planning Requirements (40 CFR 300.215)

| Planning Topics Required by OSHA | |
|---|---|
| (i) | Pre-emergency planning and coordination with outside parties |
| (ii) | Personnel roles, lines of authority, training, and communication |
| (iii) | Emergency recognition and prevention |
| (iv) | Safe distances and places of refuge |
| (v) | Site security and control |
| (vi) | Evacuation routes and procedures |
| (vii) | Decontamination |
| (viii) | Emergency medical treatment and first aid |
| (ix) | Emergency alerting and response procedures |
| (x) | Critique of response and follow-up |
| (xi) | PPE and emergency equipment |

Figure 4-3: Excerpt from 29 CFR 1910.120(q)(2) OSHA's HAZWOPER Emergency Response Plan Elements

**AGRICULTURAL GROUP PLANT SAFETY MANUAL**
**COMMUNITY EMERGENCY NOTIFICATION AND EVACUATION PROCEDURE**

1.0 PURPOSE

To provide an immediate and efficient method for evacuation of the area around the Muscatine Plant due to an emergency.

2.0 SCOPE

This procedure covers any fire, explosion, release of hazardous chemicals or any other occurrence that would affect the health and well-being of those living outside the fenced area of the plant and those traveling on the roadways.

3.0 REFERENCES

3.1 Sheriff's Department operating procedures for establishing roadblocks.

3.2 Muscatine Red Cross Emergency Shelter Procedures

4.0 DEFINITIONS

None

Figure 4-4: Excerpt from the first Pages of a Contingency Plan Used by Monsanto Agricultural Company, Muscatine, Iowa. Used with permission.

**AGRICULTURAL GROUP PLANT SAFETY MANUAL**
**COMMUNITY EMERGENCY NOTIFICATION AND EVACUATION PROCEDURE (cont.)**

5.0 RESPONSIBILITIES

5.1 The Plant Manager bears overall responsibility for all emergency actions inside the plant.

5.2 The Muscatine County Sheriff bears responsibility for establishing roadblocks around the plant area, evacuating persons who are outside the fenced area of the plant, and control at the evacuation center when established.

5.3 The Muscatine County Sheriff also includes notification to Louisa County Sheriff, Fruitland and Muscatine Fire Departments, Civil Defense, and Red Cross, in their pre-plan.

6.0 PROCEDURE (Note that this is an excerpt from a much larger document.)

6.1 In the event an emergency should occur within the plant either on the off shift or on the day shift which requires notification to homes within the 1-mile radius, notification shall be done via emergency radios and alerting siren.

6.2 When the alerting siren is sounded for notification to outside homeowners, notification shall be made to the Muscatine County Sheriff via telephone and/or radio.

6.3 Upon receiving notification of an emergency at the Plant the sheriff shall dispatch personnel to (depending on the extent) secure all roads leading into or close by the plant fenced area.

6.4 All responding personnel shall be kept informed of wind direction and any change of status of the emergency.

**PLAN A**

6.5 Should an emergency occur which does not affect personnel outside of the ½ mile boundary, the following roadblock procedure shall be followed:

1. Intersection of Wiggins Road and Ogilvie Road
2. Intersection of 57th Street and Pettibone

**PLAN B**

6.6 Roadblocks shall be established at 5 main locations, as the emergency deems necessary.

1. 57th Street and Stewart
2. 57th Street and Pettibone
3. Ogilvie Road South
4. Stewart Road South
5. Fruitland Road West

**PLAN C**

6.7 Should the emergency warrant a wider area of evacuation, the following roadblock locations shall be used. (All personnel should use extreme caution in relocating assembly points and roadblock locations.)

1. Kammerer Trailer Court
2. Acme Sand & Gravel Plant Intersection
3. Ogilvie Road South
4. Intersection of Stewart Road and G44X
5. Fruitland Baptist Church
6. Fruitland Road at Fire Station

Figure 4-4 Continued

**AGRICULTURAL GROUP PLANT SAFETY MANUAL**
**COMMUNITY EMERGENCY NOTIFICATION AND EVACUATION PROCEDURE** (cont.)

6.8 Residences/businesses within 1 mile of the plant

6.9 Emergency access gates

1. Gate 3A – Construction Gate by Sub-Station
2. Gate 7 – Entrance by the Recreation Building
3. Gate 16 – Railroad Gate
4. Gate 19 – Guardhouse (Wiggens Road)
5. Gate 8 – Fire Training Ground

6.10 Assembly points

6.10.1 Upon initial notification to personnel in the one-mile area of the plant, all persons shall proceed to the Island Church on Stewart Road. (Notice should be made of wind direction in selection of assembly points).

6.10.2 Should the need to evacuate this area become necessary, the Fruitland Baptist Church shall be used.

6.10.3 The third gathering point shall be the Kilpeck Friends Church.

6.10.4 Upon arrival at any of the three assembly points, all personnel shall notify the officer in charge (Sheriff's Dept) and report in by name.

6.11 Communication

6.11.1 The radio frequencies to be used during any plant emergency where outside assistance is requested shall be as follows.
Law Enforcement: [deleted]
Fire Department: [deleted]
Medical: [deleted]

6.11.2 Emergency control shall be maintained in the emergency control center on plant property.

6.11.3 Should the control room need to be evacuated, it shall be done via the mobile communication center.

6.11.4 The mobile center should remain as close to the plant site as safely possible.

6.11.5 Law enforcement control center shall be established at the gathering point used.

6.12 Procedure distribution

Louisa County Ambulance
Riley Ambulance
Chief Grandview Fire Department
Chief Fruitland Fire Department
Sheriff Muscatine County
Sheriff Louisa County
Chief City of Muscatine Fire Department
Director Civil Defense – Muscatine
Director Civil Defense – Louisa
Sheriff Mercer County
Rock Island County Sheriff

Figure 4-4 Continued

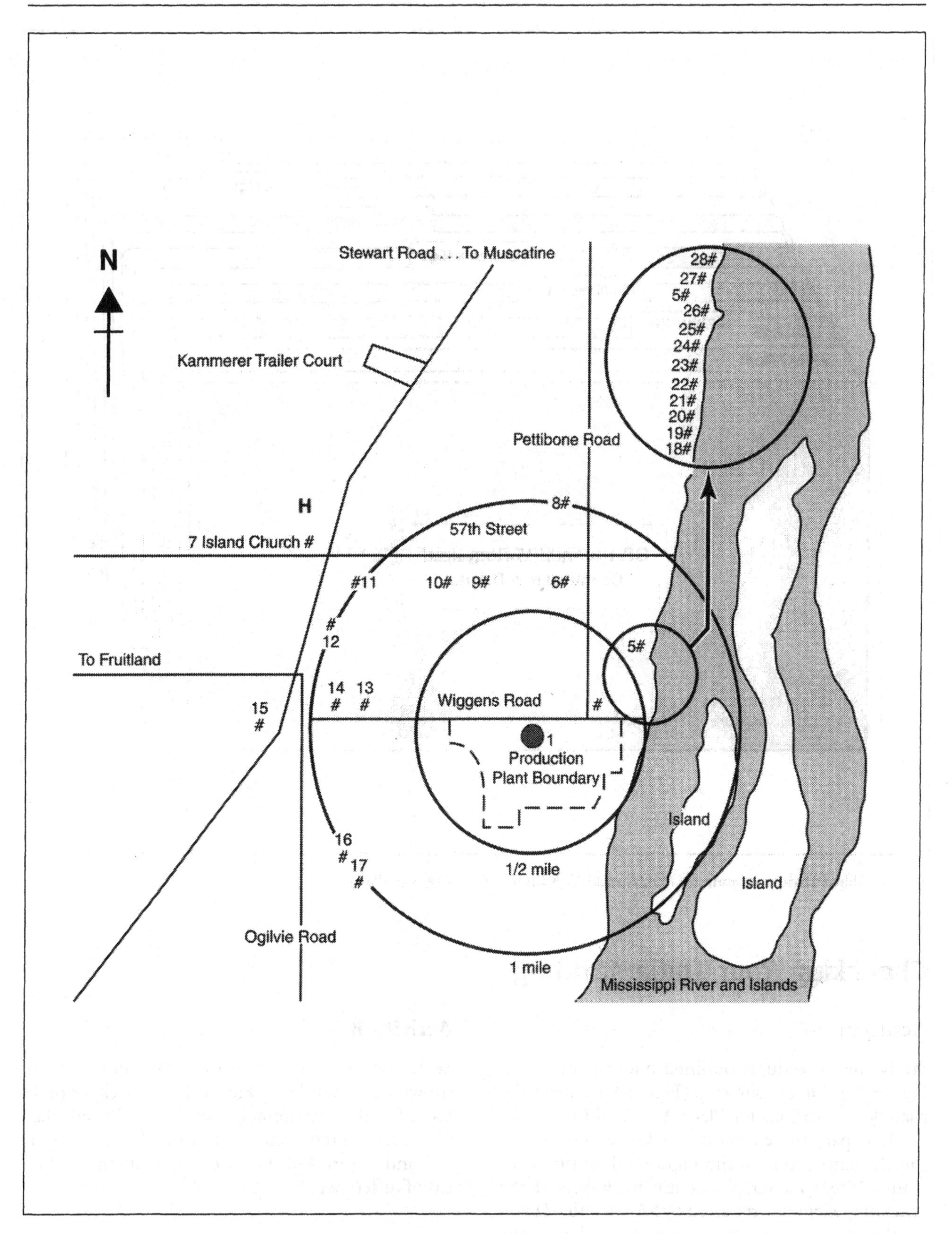


Figure 4-5: General Site Map of Monsanto's Contingency Planning Area

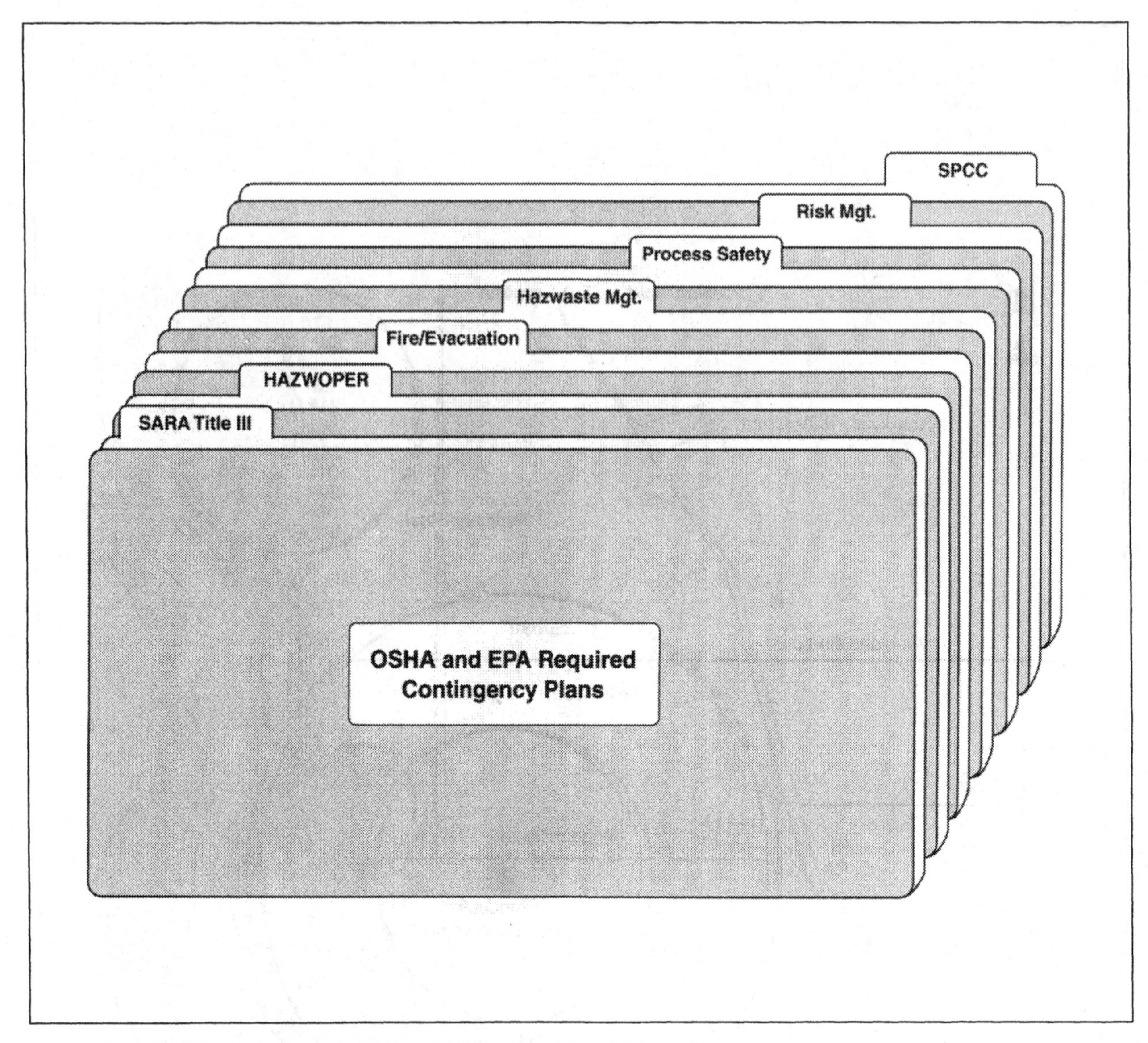


4-6: Tabbed Dividers Separating OSHA and EPA Required Contingency Plans

## Checking Your Understanding

### Activity A

Study the procedures outlined under item 6.0 in Figure 4-4. On the site map (Figure 4-5), mark the roadblock locations for Plans A, B, and C.

Compare the elements listed in Figure 4-2 to the elements found in the Figure 4-4 contingency plan. Write a paragraph identifying which of the remaining elements required by SARA Title III you would expect to find in a complete version of the contingency plan.

### Activity B

Select one of the OSHA or EPA required plans shown in the binder in Figure 4-6. Check 29 or 40 CFR for the requirements of your selected plan. Compare the requirements to those listed in Figure 4-2 and Figure 4-3. List areas of similarity and areas of difference.

## Activity C

As the environmental technician consulting with UDL (see Scenario 2 in Appendix 1), you are part of a team that is developing a new contingency plan for the company. Develop one of the planning topics required by the HAZWOPER standard emergency response (i.e., contingency) plan (see Figure 4-3). You may create or obtain any needed information as necessary. For example, if you select the topic of decontamination, use your own educational background and expertise. Or, you may obtain decontamination procedures from a reliable source, such as *the Occupational Safety and Health Guidance Manual for Hazardous Waste Site Activities*, published by NIOSH/OSHA/USCG/EPA.

## Activity D

The planning requirements of 29 CFR 1910.38(a) and (b), Employee Emergency Plans and Fire Prevention Plans, are generally applicable in any work facility. Use these requirements to analyze an existing emergency and fire prevention plan for a place with which you are affiliated (e.g., school, job, or church). Compose a list or write a paragraph that describes the omissions or areas needing improvement as required by 29 CFR 1910.38(a) and (b). Select one area that you think needs to be developed or improved the most.

Section 4-2 was adapted from materials in *Living and Writing Responsibly: Environmental Readings for Classic and Technical Composition*. Raleigh: HMTRL, 1992, by permission of the publisher.

# 4-3 Observation Reports

## Concepts

- Observation reports consist of a site inspection description and an analysis of the situation or incident.
- The objective of an observation report is to provide enough information so an informed decision can be made.
- Observation reports can be presented as a report, form, or checklist.
- Observation reports must be thorough, factual, and devoid of opinion.

Observation reports are used throughout the business world. The following is an example of a report an environmental technician might write to management regarding his observations during a visit to the company's electroplating shop:

> Upon entering the electroplating shop on 9/10/99 at 10:30 a.m., a general haziness was seen in the air and a stinging odor caused the frequent urge to cough. The haziness appeared to be concentrated in the area just above a process bath, which was bubbling vigorously. Workers removing parts racks from the process bath did not allow time for the clinging liquid to return to the bath before they were plunged into the next bath. Drag-out collectors were not used on any of the three counterflow rinse baths.
>
> All workers in the area were wearing elbow-length rubber gloves, knee-high rubber boots, and full-length rubberized aprons. It was noted that their safety glasses, fitted with plastic side-wings, and the paper dust masks covering nose and mouth were not of the OSHA-approved type for chemical splashes and mists. It was also noted that when workers attempted to talk, they frequently pulled the paper masks away from their faces using one of their chemical-laden gloved hands.
>
> Interviews were held with three workers to collect information about personal hygiene habits prior to eating, smoking, or using the restroom. A thermometer indicated that the temperature in the area was 83°F on a day when the outside temperature was 74°F. All air circulation in the 5,600 ft$^2$ process area is provided by one 1,750 cfm fan.

The purpose of an observation report is to provide accurate and objective information so informed decisions can be made, such as what modifications need to be made in the personal protective equipment (PPE) being used by the workers in the above report. Observation reports contain two basic elements: a recount of a site inspection and an analysis of a situation or incident. The methods of presenting this information can range from a traditional report to writing notes, marking appropriate items on a checklist, or filling out a form.

> Observation reports can have other names, such as inspection report, investigation report, incident report, site assessment, audit, or on-site review.

The task of the report writer is to gather information by observing the site and interviewing the persons involved. Some observation reports may be confined to a description of existing conditions; others may require a narration of events, so that a cause-and-effect relationship can be established such as in the following excerpt of a fictitious example report:

> ... the painter became so ill that doctors recommended quitting his occupation. His worsening symptoms included blurred vision, nausea, and a weakness resulting in the loss of control of his arms and legs. After months of recovery, he again contracted to paint the outside of a small building owned by Locust®. He put on clothing to cover all his bare skin, including a hood that had only a slit-like opening for his eyes. Equipped with an organic vapor respirator and gloves, the painter spent the entire day painting without experiencing any ill effects. After removing all of the protective clothing, the painter was observed by the company's environmental technician us-

ing his bare hands to methodically wash each part of the spray gun in paint thinner.

That evening, the painter became ill and vomited. When he reported to work the next day, the technician reminded him of how he had cleaned the spray gun. The painter said that this was the way he always did it and was surprised when the technician explained that organic solvents can be absorbed into the body through skin contact.

The analysis of cause-and-effect relationships may lead to the formulation of a set of recommendations, which may or may not be made by the observation report writer. In the above example, it was reported that the painter's washing method caused the second reported occurrence of adverse health effects. The obvious recommendation would be to alter the method of cleaning the equipment to prevent skin contact with the organic solvent.

The following are suggestions for preparing and conducting a site inspection.

**Ask before going:**

—Are the scope, purpose, and expectations of the inspection clearly defined?

—Do you know what process, incident, or accident site you should observe or ask questions about?

—Is there any background information you need to know beforehand, such as a manufacturing process?

—Do you know the person(s) in charge, directly involved, or most knowledgeable and have you made arrangements to talk to that person(s)?

—If you will be interviewing people, have you written a set of questions?

—Do you have the proper clothing, tools, and instruments to collect the needed information?

—Would a camera or tape recorder aid in obtaining information?

**During the inspection:**

—Objectively gather all data using careful measurements and observations.

—If possible, write down your first observations of the site before talking to anyone.

—If reviewing a process, have someone take you step-by-step through the process.

—If interviewing someone who has trouble expressing their thoughts, ask them to show you instead of tell you.

—If unsure about what someone is saying, repeat the information and ask them to correct you, expand on what they are saying, or give an example.

—After completing the interview, ask the interviewee if there is any other important information you might need to know.

**After the inspection:**

—Jot down all of your notes and impressions immediately. Don't rely on memory.

—Objectively analyze all data by comparing and contrasting and by noting patterns and similarities.

—Determine if your findings provide a sound basis on which to make a recommendation.

Although there is no standard format, formal investigative reports include some or all of the following sections:

| Section | Description |
|---|---|
| **Abstract** | Contains a summary of the report |
| **Introduction** | Contains the scope of the report, the names of the relevant persons involved, the name of the interviewer, the name of the company |
| **Materials and methods** | Details about how the information was gathered (i.e., through interviews and observations) |
| **Results** | The information that was obtained |
| **Discussion and conclusions** | Contains the cause and effect summary |
| **Recommendations** | Interpretations of the data and recommendations or conclusions based on that analysis. |
| **References** | A bibliography or list of individuals from which more information can be obtained. |

Less formal reports can be created by omitting and combining some of the above. The report in Figure 4-7 is a fictitious example of a formal on-site review. Within the report are observations, a cause-effect analysis, and recommendations. The scenario is that the writer is a waste reduction specialist from the Mississippi Waste Reduction Center who is reporting on a visit to a small metal plating shop. The intended reader or audience for the report is a shop owner who has asked for the report to help him comply with federal regulations. The purpose of the report is threefold:

—To describe the existing conditions in the metal shop

—To analyze those conditions relative to applicable federal regulations for the disposal of hazardous waste

—To make recommendations to the company for improved handling and reduction of hazardous waste

Note that none of the referenced appendices in Figure 4-7 are included.

**METAL PLATING, INC.**
**Rivertown, Mississippi**

**Waste Management:**
**On-Site Review and Recommendations**

**A. Introduction**

1. Metal Plating, Inc., located in Rivertown, Mississippi, was visited on August 14, 1999, by Jane Smith of the Mississippi Waste Reduction Center (MWRC). The purpose of the visit was to perform a waste management review focusing on wastewater from electroplating and electropolishing processes. Mr. John Doe provided the site tour and information regarding Metal Plating, Inc.'s, operations.

2. Metal Plating, Inc., rack-plates various industrial components with nickel, silver, and/or gold. A zinc plating line is available, but at the time of the review was not in service. Electropolishing of steel parts is also performed on a large volume basis. Waste management areas addressed in this report include:

   — Electropolish sludge

   — Electropolish and electroplate wastewater

3. Where appropriate, existing conditions, applicable regulations, and recommendations to assist in maintaining an effective and compliant waste management program will be addressed for each waste stream. A summary of recommendations and a general regulatory review discussing small quantity generator requirements is also included.

**B. Summary of Recommendations**

1. Generators of hazardous wastes must notify the EPA of their activities and apply for an EPA identification number. The MWRC booklet enclosed in Appendix A of the report provides the instructions and appropriate forms for obtaining an EPA identification number. This identification number must be on all documents involving hazardous waste disposal. Local and state emergency crews need to be notified of Metal Plating, Inc.'s, hazardous waste activity. This can be done by sending a letter via registered mail to the appropriate authorities. A sample letter has been enclosed in Appendix B.

2. Metal Plating, Inc., is a small quantity generator, assuming that wastewaters meet federal and local pretreatment standard prior to local sanitary sewer discharge. As a small quantity generator there are requirements concerning waste generation rates, storage time limits, and overall waste management that must be met. A detailed list of these requirements is included in Appendix C. Violation of the small quantity generator regulations may result in a fine and imposition of closure requirements for illegal operation of a Treatment, Storage, and Disposal Facility (TSDF).

Figure 4-7: Observation Report Example

**METAL PLATING, INC.**
**Rivertown, Mississippi**

**Waste Management:**
**On-Site Review and Recommendations** (cont.)

**C. Electropolish Sludge**

1. Existing Conditions:

   Metal Plating, Inc., operates an electropolishing tank for use on high-grade steel parts. Sludge is generated during this process and is removed from the tank every six months. Since start-up, total sludge generation from this process is approximately 90 gallons. All of the sludge is currently being stored in two 55-gallon, open-top, plastic drums.

2. Regulatory Review:

   Federal regulations require all waste generators to determine whether their wastes are hazardous. This may be accomplished through laboratory testing or through knowledge of the waste material or the process generating the waste. Wastes are considered hazardous due to specific EPA listing or presence of the following hazardous characteristics:

   — Ignitability

   — Toxicity

   — Corrosivity

   — Reactivity

   Although electropolishing sludge is not specifically listed as a hazardous waste, it may still be considered hazardous due to the presence of toxicity and corrosivity, two of the four hazardous characteristics. If the waste sludge is determined to be a hazardous waste, disposal must be done through a permitted facility, accompanied by a uniform hazardous waste manifest. A list of permitted TSDFs that can handle your wastes is included in Appendix D.

3. Recommendations:

   Metal Plating, Inc., should have the hazardous/nonhazardous nature of their electropolish sludge determined through testing. The appropriate tests for this determination are toxicity (TCLP) for heavy metals and corrosivity. The Mississippi State Extension CIRAS program currently has funds available for this type of testing. This service may be requested by contacting Mr. Jones at 555/555-0932. The laboratory specified by Mr. Jones should be contacted and advised of the material to be sampled and the required test parameters described in Appendix E. The laboratory should be asked to provide appropriate sampling containers and instructions to complete the sampling. If the sample data exceed the contaminant concentration limits listed in Appendix E, then the waste sludge must be treated as hazardous. If none of the Appendix E limits are exceeded, then the waste sludge is not considered by EPA to be hazardous and other disposal methods may be employed.

   For nonhazardous solid wastes, disposal in a sanitary landfill is acceptable, provided a Special Waste Authorization (SWA) from the Mississippi Department of Natural Resources (DNR) has been granted for that waste. Appendix F contains test parameters for obtaining an SWA and an application form. Please note that to obtain an SWA, there must be no free liquids in the waste. This means that the sludge should be dried prior to disposal.

**D. Electropolishing and Electroplating Wastewaters**

1. Existing Conditions:

   Metal Plating, Inc., operates silver cyanide and nickel sulfamate plating lines, as well as the steel electropolishing process. Another plating line for zinc plating is available but is not currently in service. Plated parts are allowed to hang over the plating tanks to allow as much solution dragout as possible to be returned to the plating bath; this technique also reduces contamination of the ensuing rinse waters.

   All rinses are counterflowing and are collected in a large holding tank for pH adjustment prior to disposal to the local sanitary sewer. Rinse tanks are completely emptied, cleaned out, and refilled on a weekly basis; cleaning tanks are changed over on an as-needed basis and are mixed with wastewaters for pH adjustment and eventual discharge to the sewer. Approximately 3,000 gallons of wastewater are generated per week.

   Each month an inspector from the city wastewater treatment facility takes a sample of the water between pH adjustment and discharge. Last month, analyses performed for heavy metal concentrations showed excessively high levels of all metals tested. All previous tests showed heavy metal concentrations to be well under wastewater pretreatment standards.

Figure 4-7 Continued

**METAL PLATING, INC.**
**Rivertown, Mississippi**

**Waste Management:**
**On-Site Review and Recommendations** (cont.)

2. Regulatory Review:

Wastewaters generated by jobshop electroplaters are federally regulated under 40 CFR Part 413. For electroplating and electropolishing operations that generate less than 10,000 gallons per day, the following wastewater pretreatment standards apply:

| Pollutant | Maximum Per Day |
|---|---|
| CN, Amenable to treatment | 5.0 mg/l |
| Lead | 0.6 mg/l |
| Cadmium | 1.2 mg/l |
| Total Toxic Organics (TTO) | 4.57 mg/l |

Wastewaters that exceed these limits are considered hazardous and must be treated prior to discharge to a sanitary sewer system.

State of Mississippi pretreatment standards for electroplating and electropolishing wastewaters are the same as those of the EPA. Local pretreatment standards may vary from community to community, but cannot be less stringent than those set forth by the EPA.

3 Recommendations:

The use of counterflow rinses and dragout solution recapture are excellent waste minimization practices and should continue. More frequent change out of rinse tanks and cleaning tanks will help to reduce the level of metal contamination in the wastewaters; however, once the zinc plating line is operational, it may be necessary to install a wastewater pretreatment system to comply with federal and local discharge standards. There are a variety of wastewater treatment systems to remove heavy metals such as copper, nickel, zinc, and chromium. The two main methods of treatment are (1) ion exchange and (2) precipitation with filtration.

Ion exchange involves the use of columns filled with a synthetic resin. As the wastewater is pumped through the columns, the metals are captured and retained on the resin. When the resin is full, it is emptied into a plastic bag or other appropriate container for later disposal. The columns are refilled with new resin and the system starts anew. One alternative to replacing the resin when it is full is to wash it with a mild acid solution. The resulting effluent after washing will be a concentrated heavy metal solution. This solution can then be stored for later disposal and the resin can be used again.

Precipitation with filtration involves two steps. The first step is to precipitate heavy metals from the wastewater by using a flocculent or other precipitating agent. After allowing the precipitate to settle, the water can be pumped out and discharged to the sanitary sewer. The remaining precipitate and some water are pressure-pumped through fine, plastic filters in a filter press. When the filters are full they can be scraped to remove the de-watered sludge. The de-watered sludge is then placed in an appropriate container for later disposal. The effluent from the filter press can be discharged to the sanitary sewer and the cleaned filters are reinstalled for reuse.

Information concerning these small wastewater pretreatment systems is enclosed as Appendix G.

Figure 4-7 Continued

## Checking Your Understanding

### Activity A

Conduct a detailed inspection of your kitchen, basement, garage, or workplace. Make a list of the existing safety, health, and environmental situations. Write an inspection report that lists your observations and recommendations.

### Activity B

1. Select a process about which you can write an observation report similar to the one in Figure 4-7. Analyze a part of the process or operation and then develop the following:

   —A description of the existing conditions

   —A statement of the pertinent federal environmental safety and health regulations (OSHA, EPA, DOT)

   —Recommendation(s) regarding a) improving or maintaining compliance for the process or b) reducing waste

2. Respond in writing to the following:

   —To be an effective observer or inspector, you should consider the effect your presence has on others. If you were inspecting a job site to check work practices or product quality, what effect might your activities have on workers and supervisors? Suggest an appropriate style of carrying out the inspection and what that style might imply to those around you.

   —Observation reports should be objective; therefore, you as reporter must be an objective observer, and that is sometimes a challenge. Discuss a situation where it would be difficult to be objective. What would you do to try to be objective in such a situation?

### Activity C

Read the report in Figure 4-7 and prepare written responses to the following:

—Does the language level seem appropriate for the intended reader? Make a list of the terminology that the writer presumes the reader already knows. Is that a reasonable assumption?

—Supposing that you own Metal Plating, Inc., prepare a "to do" list from the report recommendations. How well do you think the recommendation sections lend themselves to writing such a list?

Section 4-3 was adapted from materials in *Living and Writing Responsibly: Environmental Readings for Classic and Technical Composition.* Raleigh: HMTRL, 1992, by permission of the publisher.

# 4-4 SOPs and Instructions

## Concepts

- SOPs are procedures that enforce company or organizational policy.
- Most SOPs require employee training and administration to ensure, respectively, that they are understood and that they are followed and updated regularly.
- Instructions describe how to perform a task or process.
- Instructions should be tested on the people who will be using them.

SOPs and instructions are both how-to documents, but they serve different purposes. SOPs deal with policies, and instructions deal with tasks.

A SOP is typically used by the military, public service agencies such as fire and police departments, and businesses that must comply with the regulations or maintain equipment. SOP manuals set forth the steps or course of action from which workers should not deviate. For example, in working with hazardous materials, a deviation from the prescribed procedures may mean the difference between health and illness, safety and danger, and even life and death. On the job, only the most senior people in informed positions of authority can deviate from the SOP. Some organizations are so strict about employees following SOPs that failure to do so can be grounds for dismissal.

The general process for creating a SOP is as follows:

1. An authority, such as upper management, determines the need for a written policy procedure and assigns someone to write it.
2. Through a process that involves several possible steps, the writer develops the SOP. The planning steps may include but are not limited to the following:
   —Identifying the applicable regulations
   —Applying personal experience and knowledge
   —Observing involved personnel in properly carrying out the procedure
   —Studying available technologies and methodologies for carrying out the procedure
   —Questioning involved personnel about their recommendations for the procedure
3. The writer submits draft copies of the SOP to every department affected by the document and to appropriate personnel for their review. The departments must approve the document through consensus. In some instances, an employee "trial run" of the SOP may be appropriate to determine if there are any errors or problems with the procedure or writing.
4. After necessary revisions and final approval of the SOP, a designated department formally enters the document into a system or process that is similar to the following:
   —Assignment of a document number and placement of the number into a record keeping system.
   —Assignment of periodic review/revision date(s)
   —Application of management signatures
   —Placement of copies into SOP manuals
   —Distribution

Depending on the size and resources of a company, an environmental technician may be expected to do all or some of the above. Additional responsibilities may include the following:

—Developing training programs to ensure that the SOP is understood and followed correctly.

—Organizing the SOP and training manuals with related appendices (e.g., MSDSs for hazardous chemicals).

—Setting up a system for filing training dates, information about training content, lists of participants, test/checklist data, and forms signed by trainees acknowledging their participation. If a situation arises that could result in legal action against the company, a well-maintained, current recordkeeping system should provide evidence of management's intent to provide a safe workplace for employees.

—Aiding management in administering the SOP (i.e., ensuring that the procedure is followed). Employees need to know the consequences of not abiding by company procedures and need to be properly supervised.

—Establish an annual review of the procedure to determine if updates are necessary. Additionally, some work-site occurrences may reveal inadequacies, which means that re-evaluation of the procedure is necessary. Fire officials regularly debrief (i.e., review an event to gain information) after emergency incidents to learn from the event and, if necessary, refine the department's procedures.

—Obtain continual feedback from the first-line workers to whom the SOP applies.

Although there are no formal SOP format standards, it is recommended that an outline be used as in Figure 4-8. It is from the Rock Island Arsenal, which is a federal military manufacturing facility located in Illinois. As you read the SOP, keep in mind that it reflects the Arsenal's policy and, therefore, adherence is considered to be mandatory by administrators. Personnel who do not conform to the SOP could lose their jobs.

SOP 385-AO-36

**ROCK ISLAND ARSENAL**
**ARSENAL OPERATIONS DIRECTORATE**

STANDARD OPERATING PROCEDURE
No 385-AO-36 — 2 May 1999

Safety

CLEANING/REPAIR OF PLATING TANKS AND VAPOR DEGREASING PITS
AND SLUDGE REMOVAL FROM STEAM CLEANING OPERATIONS AND PAINT SPRAY BOOTH



1. **Purpose.** This Standard Operating Procedure (SOP) prescribes responsibilities and procedures for cleaning and/or repair of plating tanks, vapor degreasing pits, and removal of sludge from steam cleaning operations and paint spray booths.

2. **Applicability.** This SOP applies to all Factory Division, SMCRI-AOF, employees engaged in the operations specified in paragraph 1.

3. **References.**
   a. Rock Island Arsenal Waste Disposal Information Booklet
   b. RIAR 385-2, Occupational Safety and Health Program at Rock Island Arsenal (RIA)
   c. SOP 385-AO-33, Energy Lockout Program
   d. SOP 385-AO-38, Confined Space Entry
   e. Rock Island Arsenal Spill Plan
   f. Respirator Protection Program

4. **Responsibilities.**
   a. **SMCRI-AOF, general foreman** will:
      (1) Train employees in procedures and hazards associated with cleaning/repair or sludge removal from plating tanks, vapor degreasers, steam cleaning, and paint spray booths.

Figure 4-8: SOP Example from Rock Island Arsenal, a Federal Military Manufacturing Facility

SOP 385-AO-36

**ROCK ISLAND ARSENAL**
**ARSENAL OPERATIONS DIRECTORATE**

STANDARD OPERATING PROCEDURE

No 385-AO-36 2 May 1999

4. **Responsibilities.** (cont.)
   a. **SMCRI-AOF, general foreman** will:
      (2) Train employees in proper use of safety equipment and procedures required during cleaning/repair and sludge removal operations. This will include specific training dedicated to confined space entry (CSE) as referenced in RIAR 385-2, Chapter 27, and SOP 385-AO-38.
   b. **SMCRI-AOF, supervisor** will:
      (1) Ensure employees using a respirator participate in Respiratory Protection Program and report to Health Clinic for physical evaluation, fit test, and training if required.
      (2) Ensure all safety requirements stated in CSE permit are enforced and all required safety equipment is available and used correctly.
      (3) Ensure all requirements for CSE (reference RIAR 385-2, Chapter 27, and SOP 385-AO-38) are enforced. This is to include the use of portable monitors for explosive atmospheres, oxygen deficiency, and carbon monoxide.
   c. **SMCRI-AOF, worker** will:
      (1) Use approved safety equipment correctly as indicated on CSE permit.
      (2) Use required respirator equipment when entering tank or pit as prescribed on CSE permit.
      (3) Know and comply with all safety requirements applicable to the operation.
      (4) Immediately report any hazardous substance spill to the supervisor, as noted in RIA Spill Plan.
      (5) In the event of an emergency, send for help at once. Immediately notify RIA Fire Department, extension 117, to assist in handling the emergency situation. **DO NOT ENTER THE CONFINED SPACE.**

5. **Procedures.**
   **SMCRI-AOF, supervisor** will:
   a. Contact Industrial Hygiene Office, HSXP-RIA, extension 20807, or Safety Office, SMCRI-SF, extension 21381, to give notice of intended operation, request atmosphere testing and issuance of CSE permit. Contact will be made one day prior to beginning operation as per SOP 385-AO-38.
   b. Contact RIA Fire Department, extension 25948, to obtain fire permit if repair operations involve combustible materials or hot work to include welding, cutting, or torching.
   c. Contact Lubrication and Maintenance Support Unit, SMCRI-AOF-RSL, to request equipment (air line respirator, safety harness, gloves, boots, coveralls, etc.). Air line respirator equipment must be requested at least one day in advance.
      **Note:** Respirator and scrubber board will be returned immediately after completion of operation for sanitizing before reissuing.
   d. Place liquid/sludge waste in drums and label accurately. Provide turn-in documents for disposal of waste in accordance with Waste Disposal Information Booklet.
   e. Ensure employees entering paint spray booth pit or water wash booth pit for any reason or entering a vapor degreasing tank for repair/cleaning operations use protective equipment as stated in CSE permit.
   f. If sludge removal or repair work requires an employee to enter a tank (as stated on CSE permit), purge remaining vapor with forced air blower fans for a minimum of 30 minutes.
   g. Ensure at least one employee is at tank side to constantly monitor each employee in the tank.
   h. Lock out or tag out utilities to the tank as necessary and monitor operation.
   i. After operation is concluded, return used equipment to issuing office.
   j. Encourage employees involved in repair/cleaning/sludge removal to shower after operation is complete.
   k. If CSE permit expires prior to operation completion, contact SMCRI-SF or HSXP-RIA for issuance of a new permit

The proponent of this SOP is Industrial Management Office, SMCRI-AOI.

Signature: ______________________________
Director, Arsenal Operations Directorate

Distribution:

SMCRI-AOF (Foremen/Supervisors)
SMCRI-SF
SMCRI-SEM (Environmental Coordinator)
HSXP-RIA

Figure 4-8 Continued

## Instructions

Instructions are a common part of everyday life, ranging from a two-page leaflet that comes with a toaster to a four-inch thick software manual. As an environmental technician you may be asked to write instructions for a wide variety of tasks, such as how to calibrate a combustible gas meter, prepare and handle sampling containers, decontaminate sampling equipment, or dispose of hazardous wastes, to name just a few.

The writing of instructions should begin with questions similar to the following:

—Who will be using the document?

—Do they really need documentation? Will another means of communication work better than written instructions, such as stickers like the ones used in photocopy machines that illustrate how to open various drawers?

—When will the reader be using the document?

—Why will they be using the document?

—Is it appropriate to obtain workers' input while developing the document or to ask for their review or approval before finalizing the document?

—Will the reader need an explanation for why the document is taking effect?

—Will the reader need training in connection with implementation of this document?

When writing instructions follow the basic writing steps and rules described in Chapter 2. Note that illustrations and white space are particularly useful in instruction documents. White space is the area on the page that does not have text. Use white space generously as it is difficult to read step-by-step instructions on a page crowded with text.

## Instruction Format

Instruction formats vary as much as the tasks that they describe. There are, however, some basic components, which include the following:

—Title

—Date

—(Optional) Document number and revision number

—An introduction or overview of the task

—(Optional) List of terms

—(Optional) List of equipment or resources needed to perform the task

—The steps involved in completing the task, with each step containing the following:
- Description of the step
- (Optional) Example
- (Optional) Additional information relevant to performing the step

Figure 4-9 contains information about how to write effective instructions. As you read, think not only about the message, but also about how the layout of the material contributes to your understanding.

| WRITING EFFECTIVE INSTRUCTIONS | |
|---|---|
| Page number: | (most clearly expressed as page 1 of 5, page 2 of 5, etc.) |
| Document #: | (for record keeping purposes) |
| Written by: | |
| Date prepared: | (date the final version was completed) |
| Approved by: | (for administrative purposes) |
| Next review date: | (to ensure document is kept current) |

Figure 4-9: Writing Effective Instructions

**Introduction**

"Writing Effective Instructions" is designed for use by technicians in the environmental technology field who must write operational instructions for first-line workers and supervisors.

The definition of instructions is a written communication that tells the reader how to perform a task.

The purpose of this material is to teach the reader how to write instructions. This document consists of four major subsections:

1. Plan Your Instructions
2. Write the First Draft
3. Test and Revise the Instructions
4. Implement the Instructions

**Terminology**

| | |
|---|---|
| Diction | The writer's selection of wording. |
| Operator | The experienced worker who reads and follows the instruction document as the writer observes. |
| Task analysis | A detailed, orderly listing of what a worker does and knows to perform a skill. |
| Trainee | The inexperienced worker who learns how to perform the task by reading the instruction document. |
| Verb | A word that represents an action, such as "sing" or "push." |

**Equipment, Materials, and Resources**

— Computer with desktop publishing program

— Experienced worker

— Audio or video recorder

— Audio cassettes or video cassettes

**Steps**

**1. Plan Your Instructions**

1.1 Identify your purpose.
Is your purpose to lead an operator through a series of steps, or is it also to educate the operator as to why processes occur as they do? Suppose, for example, your purpose is to post at a work station the instructions for a particular emergency procedure. This will affect the format and language of your instructions (i.e., emergency instructions should be formatted in large, bold print, and written as commands in as few words as possible).

1.2 Identify readers' skills.
Plan a level of language (i.e., diction and sentence structure) appropriate for the intended user of your instructions.

Figure 4-9 Continued

**Steps (cont.)**

**1. Plan Your Instructions (cont.)**

1.3 Perform a task analysis.
Select an experienced operator to perform the steps while you take notes or tape what you observe. While observing and writing, ask the operator why she did the steps. Ask "why" questions during a task analysis to help you determine what the operator knows (that is, the operator's thought processes that contribute to performing the steps properly) relative to what the operator does. Include those thought processes in the list of observations, so the points of operator knowledge will be in the final draft of the instructions.

1.4 Design graphic aids.
Determine if graphic aids like charts, graphs, or illustrations will contribute to the reader's understanding of the instructions.

1.5 Organize headings.
If many steps result from the task analysis, keep your reader from being overwhelmed by grouping the steps. Write an appropriate heading for each grouping; for example, a first heading might be "Preparation."

**2. Write the First Draft**

2.1 Compose the introduction to include the following:
— Procedure name and definition
— Purpose of procedure
— Intended users/readers
— Major divisions of instructions, if any

2.2 List needed equipment, materials, and resources.

2.3 Define technical terms.

2.4 State warnings/cautions/safety requirements.
Warnings and cautions may be stated at the beginning of the document or just before the appropriate steps in the instructions. Usually in technical writing, the use of warnings is reserved to alert the user about safety issues. Include informational notes whenever appropriate.

**CAUTION:** When choosing action verbs, be aware of references to complex concepts. For example, in the sentence "Plug the hole in the hazardous materials container," the word plug is a complex action because substeps are involved in plugging. The substeps may need to be listed separately if they are not already generally known by the trainee. Examples of complex action steps that require substeps include:

— Remove the lid from the drum of hazardous waste
— Decontaminate the suit
— Neutralize the chemical
— Take a sample

2.5 Write the steps.
Typically, instructions are written as a series of numbered commands, with each command containing one verb; for example, instructions for operating a certain fire extinguisher are as follows (the verbs are underlined):

1. Pull ring pin.
2. Start from eight feet away from fire.
3. Aim at base of fire, and keep extinguisher upright.
4. Squeeze lever, and sweep from side to side.

*Note:* If steps are complicated, divide them into substeps that are easy to follow. You might place letters before each substep or use decimal numbering, for example, (a), (b), etc. or 1.1, 1.2, etc.

Figure 4-9 Continued

**Steps (cont.)**

**2. Write the First Draft (cont.)**

2.6 Submit draft for review.
Review the completed first draft with the operator. Next, test the steps; that is, walk through them, doing the actions in order, constantly asking if you have included every necessary substep and thought process. Be aware of unclear steps or possible omissions that may confuse an inexperienced employee or even keep him/her from being able to proceed beyond a step. Retest and revise instructions until you are satisfied that they are both complete and understandable.

**3. Test and Revise the Instructions.**

3.1 Test the instructions on a trainee who is not experienced in the operation.
Select a trainee who represents the type of worker that will be a typical user of the instructions.

**WARNING:** Weaknesses in your test instructions could endanger an inexperienced operator/trainee or damage expensive industrial equipment. An experienced operator must be present if these are possibilities.

3.2 Identify weaknesses in the document.
Watch for hesitation points by the trainee. Facial expressions may provide clues to points of confusion. Determine at the conclusion of the test if the results of the trainee's performance meet the job requirements of safety, productivity, and quality.

3.3 Revise the draft to correct weaknesses.

3.4 Re-test the instructions.
With your manager and/or the experienced operator, analyze the advisability of re-testing with the original trainee or someone else. Note that revision and re-testing may need to be repeated numerous times.

3.5 Compose final draft.
Many word processing programs, especially those with desktop publishing capability, are useful for creating readable layouts.

3.6 Submit draft to manager for final approval.

**4. Implement the Instructions**

4.1 Make instructions available to appropriate personnel.

4.2 Document training in employee records.
Keep attendance lists and performance checklists in the company records. Some companies ask employees to sign and date a printed statement acknowledging that they have read the document.

4.3 Request feedback.
Suggestions for new materials and for more innovative, efficient, and cost-effective methods often arise from first-line workers.

4.4 Expect managers to handle instruction compliance among their employees.
Employees who do not follow instructions may endanger themselves, their coworkers, and company equipment.

Figure 4-9 Continued

## Checking Your Understanding

### Activity A

Write a paragraph regarding the purpose of the SOP in Figure 4-8. Is the purpose of the document to:

1. Express the Rock Island Arsenal's policy regarding the procedure
2. Tell the employees how to clean/repair plating tanks, vapor degreasing pits, etc.
3. Fulfill both 1 and 2

One way to determine an answer is to ask, "If I were a new employee inexperienced in cleaning and repairing plating tanks, etc., could I correctly complete the procedure by reading the document?"

### Activity B

With the assistance of your teacher or work supervisor, select a topic (e.g., the proper maintenance of respirators), conduct the research necessary to inform yourself of the regulations, and write a SOP for the process.

### Activity C

(See Scenarios 1 and 2 in Appendix 1.) Write a SOP that informs Locust® or UDL employees of company policy regarding one of the topics below. Incorporate requirements from OSHA's General Industry Standards (29 CFR Part 1910) as necessary.

—Waste oil management

—Solvent management

—Safety glasses

—Hearing protection

—Machine safety guards

—Food consumption

—Waste disposal

—MSDS recordkeeping and use

### Activity D

1. Identify a task done at school or home that, if not properly performed, poses potential problems to health and safety, the environment, or equipment maintenance. Choose a task that does not already have written instructions. Write step-by-step instructions for the process, following the instructions format in this section. Suggested tasks include:

   — Evacuation from classroom (or home) in event of emergency

   — Disposal of school (or household) waste materials

   — Equipment shutdown for maintenance/cleaning (e.g., overhead projector at school, coffee maker at home)

   — Small equipment safety (e.g., coffee maker, blender, lawnmower)

   — Everyday procedures (e.g., make a sandwich, set the table, change a light bulb)

2. Ask a classmate to test and critique your written instructions. Together, evaluate the effectiveness of the instructions. Revise the instructions, incorporating the suggested changes, and ask the classmate to review the changes.

### Activity E

Using an MSDS, such as the MSDS for Stoddard Solvent in Appendix 4 or an MSDS found on the Internet, write a set of instructions. For example, you could give instructions for how to use solvent to clean a spray gun and include information about storage, handling, spill response, fire response, and personal protection from exposure.

# 4-5 Logbooks and Field Notes

## Concepts

- Logbooks contain tables of data, such as meter readings.
- Field notes are journals containing documentation of events and observations.
- A field notebook can be a useful personal record and a valuable source of information for others who happen to take over a job or project.
- Logbooks and field notes prevent the repetition of completed tasks and provide important legal documentation.

Logbooks and field notes are used to document and create a permanent record of information, notes, and activities. Some documentation takes place in the field; some is more appropriate to the office.

Logbooks contain records of specific scientific data required by regulation or company policy, such as emission levels from a smokestack. They are usually filled with tables or columns of data written in pen that are dated and initialed. Field notebooks contain narratives describing the events of a day's fieldwork.

In both books, the documentation must be clear and understandable, so colleagues can read the information at a later time. One of the most frequent causes of financial loss for a firm is inadequate documentation, which results in the necessity of repeating work already done. In addition, carefully written and detailed documentation can become valuable evidence in legal proceedings.

## Logbooks

A simple example of a logbook is the sign-in sheet that many companies require visitors to sign when entering and leaving company premises. For the most part, the data recorded in logbooks is straightforward. Since a logbook is a required type of recordkeeping, companies normally have a SOP regarding the type of information that must be collected and how it must be recorded. Figure 4-10 shows the requirements for a logbook from an EPA publication.

"Sample control procedures are necessary in the laboratory from the time of sample receipt to the time the sample is discarded. The following [procedure is] recommended for the laboratory:

The custodian [of the sample] must maintain a permanent logbook to record, for each sample, the person delivering the sample, the person receiving the sample, date and time received, source of sample, date the sample was taken, sample identification log number, how transmitted to the laboratory, and condition received (sealed, unsealed, broken container, or other pertinent remarks). This log should also show the movement of each sample within the laboratory; i.e., who removed the sample from the custody area, when it was removed, when it was returned, and when it was destroyed. A standardized format should be established for logbook entries."

(Source: *Manual for the Certification of Laboratories Analyzing Drinking Water: Criteria and Procedures Quality Assurance*, EPA 815-B-97-001 (http://www.epa.gov/OGWDW/certlab/labindex.html), U.S. EPA, Office of Ground Water and Drinking Water, 1997.)

Figure 4-10: Example of Requirements for Logbook Entry

Figure 4-11 illustrates the types of information that is typically collected in an analytical laboratory's logbook. An analytical laboratory is a laboratory certified to perform analysis of hazardous samples.

## Field Notes

Field notebook entries often include the following:

—A narration of the events in a day's field work

—Data collected

—Personal observations and evaluations of conditions and situations

—Sketches that illustrate and clarify events and observations

| LOGBOOK | | | | | | | |
|---|---|---|---|---|---|---|---|
| Analyte Laboratories, Inc. | | | | | | | Riverside, CA |
| (Please print clearly) | | | | | | | |
| Sample Description: | | | | Sample ID No.: | | | |
| Initial Receipt | | | | | | | |
| Delivered By | Received By | Date Received | Time Received | Sample Source | Date Taken | Method of Transmittal | Condition* |
| | | | | | | | |

| In-House Handling | | | |
|---|---|---|---|
| Name | Date Taken | Date Returned | Date Destroyed |
| | | | |
| | | | |
| | | | |
| | | | |
| | | | |
| | | | |
| | | | |
| | | | |
| | | | |
| | | | |
| | | | |
| | | | |
| | | | |
| | | | |
| | | | |
| | | | |
| | | | |
| | | | |

*Sealed, unsealed, broken container, etc.

Figure 4-11: Example of a Logbook Page

Site 236 Page 1 of 16
June 8, 1998

Weather - 75°F; high expected @ 88°F
Rain/scattered showers in morning, clearing this afternoon.

Scope = Remove (2) 6,000 gallon storage tanks - gasoline
(1) 500 gallon used oil tank
(1) dispenser island
All tanks to be < 3" product in tanks and in single basin.

7:00 a.m. - Site meeting
Present: [XYZ] Excavating Co.
Steve Brown
Bob Long
Site manager
Doug Polk
General contractor
John Connor

Discuss health and safety procedures. Discuss route to hospital and confirm emergency numbers with Doug. Doug indicated that 5th Street closed, needed to take 6th Street detour to United Medical Center (approx. 10 min. drive).

[Notes continue in the original but are omitted here.]

Site 236 Page 7 of 16
June 8, 1998

12:30 - 1:00 pm
Break for lunch

1:00 - 1:30 pm
Remove concrete from gasoline storage tank area. Load truck and Bob transports to landfill.
Bob gone 1:30 - 2:15 pm. 45 min round trip.

1:35 pm
Sand for backfill arrives - first of the loads ordered. A&M Sand and Gravel Co.

1:40 pm
Pump removal from gasoline island. Red Jacket pumps nice. Unfortunately plumbed with mix of copper and rubber flex lines. No longer to code. Surface staining noted at pipe fittings.
Suction system; but rubber hose set at low stop - looks like gasoline collects in area and deteriorating pipe.
Surface staining appears limited to 3' around line - would be much worse in a couple of years.
See photograph.

[Notes continue in the original but are omitted here.]

(Source: John Gaines; Forest Road Consulting, Inc.)

Figure 4-12: Field Notes Example

A rule of thumb for what field notes should contain is all of the information a colleague would need to take over a job or project. When writing field notes it is good to ask periodically "If I were to quit today, could a colleague use my field notes to continue my job tomorrow?" As with logbooks, carefully written, detailed field notes can become valuable evidence in legal proceedings.

The following are general suggestions for making entries in field notebooks:

—Follow the company's SOP for documentation or, in its absence, ask about your employer's expectations regarding field notebook entries.

—Write neatly so that all data, especially numbers, can be read accurately.

—For field communications, some employers expect entries to be written with waterproof ink for permanence. To alter data, draw a single neat line through the information to be changed and enter the correct information. Write your initials by the correction.

—When decisions must be made in the field that differ from a plan (such as a sampling plan), record your rationale.

—When describing a procedure, ask if your written and/or sketched information would enable someone else to reproduce each action.

—For accuracy and completeness, write your entries as they occur. Do not complete field notebook entries back at the office or at the end of the day.

—Date and initial each entry.

Figure 4-12 shows actual field notes written by a worker for an environmental consulting firm.

## Photographs and Sketches

Photographs are usually the best media for documenting important scenes or points of interest because they provide the most accurate visual record. At times, however, photos aren't possible or even desirable. For instance, in rugged terrain camera equipment could be too much of a burden to take along. In such instances the other alternative is to make a sketch in a field notebook. The following are basic steps for making renderings clearer and easier to understand:

1. Determine the basic shape(s) and how it should be modified to fit what you see; for example, the shape may be flattened or lengthened. (See the section in Chapter 3 on graphics.)

2. Draw a north arrow and orient the sketch to the north.

3. Sketch the scene. It is not always necessary to create a precise scale drawing for sketches. Where measurements need to be indicated, write the numerals and units clearly.

4. Label your sketch. If needed, write brief descriptive details on the sketch. If descriptions are lengthy, identify them with "A, B, C" labels, and then below the sketch, write the appropriate material corresponding to those labels.

5. If site measurements are necessary, determine a permanent site feature as a starting point for your measurements. Measure as accurately as required by the regulations or by plan specifications.

## Checking Your Understanding

### Activity A

During a day at home or at work, keep "field notes" of events, decision-making, data, and so on. Include the following entries:

—Narration and description of events, decision-making, etc., through an eight-hour day. Note the time for each activity (approximate 15-minute increments).

—Draw at least one sketch that is complex enough to combine several shapes.

—Take at least one set of measurements.

—Use one page to create a table on which you will collect simple data for one type of information that is likely to keep occurring during the day (e.g., phone ringing, cars passing, etc.)

—Print and draw in ink.

—Sign and date the record.

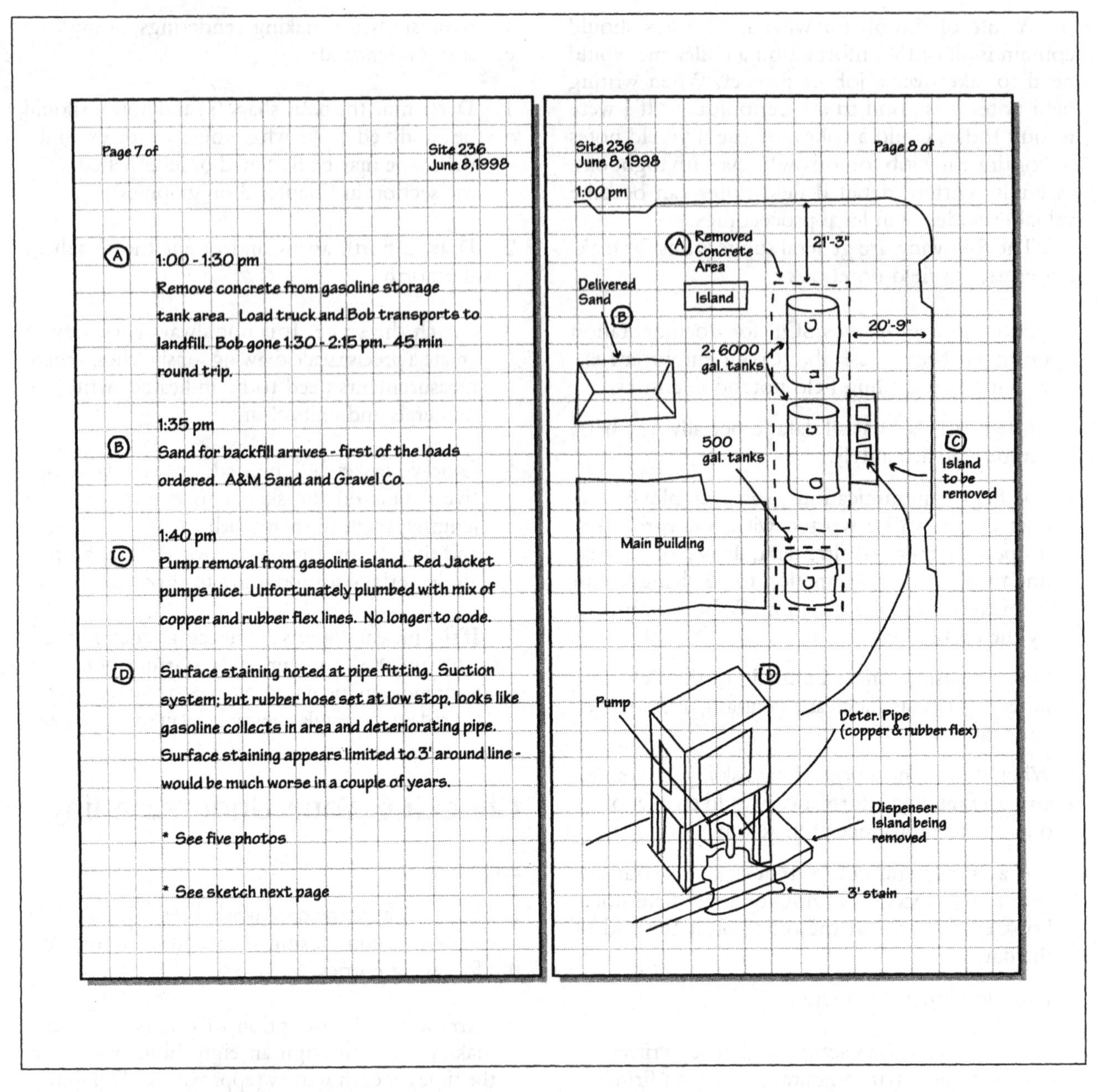


Figure 4-13: Field Book Example with Grid Pages

You may simulate field notebook pages by using graph paper (see Figure 4-13) or use an actual field notebook. If using graph paper employ left and right pages to simulate a book

Your instructor will be assessing:

—Your ability to communicate your day in such a way that he can visualize your activities

—Legibility

—Quality of your sketching and labeling

—Evidence that your entire day is accounted for

# Summary

Depending on the company or organization, environmental technicians can be expected to write technical documents such as contingency plans, observation reports, SOPs, instructions, field notes, and logbook entries.

Contingency plans are the procedures required by law in the event of emergencies or accidents. OSHA and EPA require numerous written contingency plans. The general method of writing contingency plans is to write a thorough basic contingency plan based on OSHA and EPS requirements, put it in a binder for accessibility and training purposes, and add contingency plans as needed, separating plans with tabbed dividers.

The purpose of the observation report is to communicate the results of a visit to a particular site and to analyze a situation or incident. Accurate and objective writing enables readers of the report to make informed decisions. Observation report methods involve describing observable conditions or narrating events to establish cause-and-effect relationships, which lead to the making of recommendations. The writer of an observation report may analyze cause-effect and make recommendations, or simply gather data and forward the information to others for analysis and evaluation.

SOPs reflect the policy of an organization, and adherence is usually mandatory. The writing of a SOP is a collaborative effort, where written drafts must be reviewed and approved by many managers. In addition to writing a SOP, an environmental technician may be expected to develop training and a document administration procedure.

Writing effective instructions involves studying the audience and how a particular task is performed. In addition, instructions should be pre-tested by the employees who will use them.

Logbooks contain tables of recorded data. Field notebooks contain a worker's daily record of data, activities, and the thinking behind decisions. Entries consist of text and simple diagrams that document an environmental technician's work for personal reference, for colleagues taking over the project, and for legal purposes.

PRESERVING THE LEGACY

5

# Environmental Compliance Forms

## Chapter Objectives

Upon completing this chapter, the student will be able to:

1. **Demonstrate** the accurate and legible completion of forms.
2. **Explain** the responsibilities associated with certification signatures.
3. **Locate and submit** the correct form to the appropriate agency.

## Chapter Sections

# 5-1 Completing Forms

## Concepts

- Incomplete or inaccurate forms can cost employers thousands dollars and even put people's health or lives in danger.
- Forms must be legible and either typed or written in ink.
- Certified signatures are affidavits swearing that the contents of a form are true.
- Certified signatures are usually provided by professional engineers, owner-operators, or upper-level managers.

In addition to writing letters and technical documentation, an environmental technician's responsibilities usually include completing a variety of government forms. Most of these forms are required by federal and state regulatory agencies and by local air, fire, water, and sanitation districts. The different kinds of forms that environmental technicians are required to complete can number in the dozens, depending on their job responsibilities, the nature of their company's business, and regional and state regulations. The ability to complete these forms accurately and legibly is a critical skill to employers. The reason is that an incorrectly completed form could cost a business thousands of dollars in fines or someone their life.

Missing or inaccurate information can be deadly. For example, a Uniform Hazardous Waste Manifest is a required form that accompanies all hazardous waste in transport. Should an accident occur during shipment, the form contains crucial information needed by emergency response personnel for effective incident management. The manifest is one of the tools used by the first responder to determine what immediate course of action should be taken to decrease the hazard. The government takes the accuracy of this form very seriously and can levy fines for mistakes.

Another example of a form considered critical by regulating authorities is the Material Safety Data Sheet (MSDS). An MSDS contains specific chemical information regarding a substance's composition. Companies are required to have copies of MSDSs for all hazardous substances on site and to make these data sheets available to employees. Although the manufacturer or supplier of a product is required to provide an MSDS, the technician should assume the responsibility of verifying that it is current and complete. Since an MSDS also contains important health and safety information, penalties can be levied for having incomplete or inaccurate information.

Since mistakes are costly and the number of required forms continues to increase, an environmental technician should to know how to complete forms accurately and efficiently.

"For such simple mistakes as no customer phone number listed, or a signature in the wrong place, fines typically range from $50 to $1,000, and can go as high as $25,000! The more accumulated errors...the stiffer the penalties become."

(Source: Safety-Kleen Corp., *Oil/Vacuum Services Specialist*, Elgin, IL: Advanced Certification Training, Safety-Kleen Corp., 1997.)

The EPA recently published in the *Federal Register* a proposed rule that would require more reporting to the Toxics Release Inventory (TRI). The intent is to lower the reporting thresholds for 13 chemicals or chemical groups to either 10 lb or 100 lb per year, depending on the chemical's persistence in the environment and bioaccumulative capacity. If this proposed rule becomes final, it will affect a wide range of industries and the EPA estimates that the extra reporting mandated by the proposal would require facilities to submit a total of 17,000 additional reports annually. The EPA further estimates that this would result in total industry costs of about $126 million in the first year and $70 million each year thereafter.

Reprinted from Hileman, Bette, *Chemical & Engineering News*, Volume 77, Number 2 (January 11, 1999), p. 4, with permission. Copyright © 1999 American Chemical Society.

## Accuracy and Completeness

Safety-Kleen Corporation, a national waste management firm, regularly deals with small quantity generator (SQG) customers' manifests and is well acquainted with their customers' concerns. Safety-Kleen estimates that one out of every four completed manifests contains errors.

It is true that filling out forms can be monotonous, and monotony can lead to carelessness. To reduce the number of inaccurate or incomplete documents, here are some commonsense steps to follow:

—Make sure that the form you are using is the latest version.

—Unless otherwise indicated, type or fill out the form in ink. If possible, obtain the form in an electronic format, and print it using a laser printer or return it via e-mail.

—Keep a dictionary handy and use it whenever you are in doubt about spelling, word choice, etc.

—Keep a reference file of copies of similar forms that have been completed. If you have a question about the information required on a form, it may help to refer to a previous form of the same type.

—For complex forms, make a copy, fill out the copy as a sample, double-check it for accuracy, and then carefully transfer the information to the form to be submitted.

—After completing a form, always double-check your work.

> "Don't worry about wasting someone's time by asking questions. It's far better to ask a few questions than to do your work wrong. I think I'd rather work with someone who's annoying but diligent, than with someone who's pleasant but incompetent!"
>
> Sara Bartholomew
> Quality Assurance Supervisor for
> Technical Publications Nomura Enterprise Inc.

## Legibility

Accurate and complete information is useless if no one can read it. One federal agency, the U.S. Department of Transportation, agrees and counts an illegible entry on a manifest as an error. Companies can be fined for legibility errors, particularly when they occur on a form containing health and safety information.

The ability to produce legible documents is partially dependent on good judgement and work habits. The following are some practical steps that can be used to ensure the completion of legible documents.

—Whenever possible type information rather than write it by hand. Depending on the company's preferences and the availability of equipment, the technician might find it useful and more efficient to fill out forms electronically. In most cases, it is well worth the initial time and cost to obtain or create standard forms for computer use. The individual form can be printed out for hard copy use, and its field information input into a database that tracks the paperwork (i.e., when forms were sent or received, if the form is being sent a second time, etc.).

—Use print (rather than cursive) letters when filling out a form, except for places where a signature is required.

—On forms with attached duplicate copies, type or press hard to ensure all copies are legible.

—Whenever possible, try to fill out forms when you are not in a hurry. It might help to set aside a block of time during the day or week for completing forms.

—When filling in numbers on a form, use consistent, standard characters. For instance, in the U.S., it is usually preferable to write the number seven without a European slash (i.e., 7), because of the possibility of mistaking it for another character. The number zero is another possible source of confusion. If you are writing by hand, a slash through a zero (i.e., Ø) will help distinguish it as a number rather than a letter. If you type the forms, however, this key will probably not be on the typewriter. In this situation, the technician should check with a supervisor to see if there is a preference for the method of representation.

## Certifying Signatures

Some forms require a certified signature. Certification is defined as follows:

> "By definition, the words certify, warrant, or guarantee mean to assure the total accuracy of something or to confirm that a standard has been met absolutely. Legally, these words and their derivatives are virtually interchangeable. Therefore, if you certify or warrant something, you have sworn that something is unequivocally true or correct or perfect."[1]

Figure 5-1 is an example of a certification statement from EPA Form 8700-12, *Notification of Regulated Waste Activity.* Hazardous waste generators must file the notification form to obtain an EPA identification number for their hazardous waste record keeping activities.

In most cases, the responsibility of signing a certification should rest with a professional engineer, an owner-operator, or an upper-level manager. In some forms this is explicitly stated. For example, the line-by-line instructions that accompany EPA Form 8700-12 state that the "certification must be signed by the owner, operator, or an authorized representative of your installation." An "authorized" representative is a person responsible for the overall operation of the business (i.e., a plant manager or superintendent, or a person of equal responsibility).

[1] Sheila A. Dixon and Richard D. Crowell, *The Contract Guide*, Monterey, CA: DPIC Companies, Inc., 1993.

Certification: I certify under penalty of law that this document and all attachments were prepared under my direction or supervision in accordance with a system designed to assure that qualified personnel properly gather and evaluate the information submitted. Based on my inquiry of the person or persons who manage the system, or those persons directly responsible for gathering the information, the information submitted is, to the best of my knowledge and belief, true, accurate, and complete. I am aware that there are significant penalties for submitting false information, including the possibility of fine and imprisonment for knowing violations.

Signature ____________________

Figure 5-1: Certification Statement Example

Technicians who gather data on behalf of an engineer or manager should be able to verify their work with documentation about how data was collected and recorded. In addition, it is recommended that the technician walk the responsible manager or engineer through the information gathering and reporting processes, so he is personally familiar with how data was obtained. If a manager or engineer waives responsibility of these verification efforts, then the technician should document that as well.

> "The responsibility of signing a certification should be at the professional engineer (PE) level, the owner-operator level, or the upper manager level."
>
> Sally Gaines
> Environmental Compliance and Technology
> Instructor Scott Community College

Guidance on liability issues may be obtained from a company's attorney or insurance carrier. Most insurance companies provide information services regarding the correct completion of government forms to prevent clients from becoming liable for damages covered by their policy.

## Checking Your Understanding

### Activity A

As a class or in smaller groups, simulate a hazardous materials transportation accident. The simulation could involve a "hazardous material" spill (flour, corn meal, or vegetable oil), a parked truck bearing actual or simulated placarding, and perhaps even dry ice vapor coming from within the truck. Extra "acting" may even be arranged to go with the scene. The more complex and creative the simulation, the more challenging the activity will be. Each student must observe the simulation and do the following:

1. Accurately fill out sections I, II, and IX of the blank DOT Hazardous Materials Incident Report (see Figure 5-2) using personal observations and information obtained from interviews with the actor(s). Note that, depending on the simulation, other sections of the form can be completed.

DEPARTMENT OF TRANSPORTATION

## HAZARDOUS MATERIALS INCIDENT REPORT

REQUIREMENTS: The regulations requiring reporting of hazardous materials incidents are contained in the Code of Federal Regulations (CFR), Title 49 Parts 100 to 179 (governing the transport of hazardous materials by rail, air, water and highway). Failure to comply with the reporting requirements contained therein can result in a civil penalty.

A Guide for Preparing the Hazardous Materials Incident Report is available from the Information Systems Manager, Office of Hazardous Materials Transportation, DHM-63, Research and Special Programs Administration, U.S. Department of Transportation, Washington, DC 20590.

PUBLIC REPORTING BURDEN FOR THIS COLLECTION OF INFORMATION IS ESTIMATED TO AVERAGE 1 HOUR PER RESPONSE, INCLUDING THE TIME FOR REVIEWING INSTRUCTIONS, SEARCHING EXISTING DATA SOURCES, GATHERING AND MAINTAINING THE DATA NEEDED, AND COMPLETING AND REVIEWING THE COLLECTION OF INFORMATION. SEND COMMENTS REGARDING THIS BURDEN ESTIMATE OR ANY OTHER ASPECT OF THIS COLLECTION OF INFORMATION, INCLUDING SUGGESTIONS FOR REDUCING THIS BURDEN, TO INFORMATION SYSTEMS MANAGER, OFFICE OF HAZARDOUS MATERIALS TRANSPORTATION, DHM-63, RESEARCH AND SPECIAL PROGRAMS ADMINISTRATION, U.S. DEPARTMENT OF TRANSPORTATION, WASHINGTON, DC 20590: AND TO THE OFFICE OF INFORMATION AND REGULATORY AFFAIRS, OFFICE OF MANAGEMENT AND BUDGET, WASHINGTON, DC 20503.

**DEPARTMENT OF TRANSPORTATION** **HAZARDOUS MATERIALS INCIDENT REPORT** Form approved OMB No. 2137-0039

INSTRUCTIONS: Submit this report in duplicate to the Information Systems Manager, Office of Hazardous Materials Transportation, DHM-63, Research and Special Programs Administration, U.S. Department of Transportation, Washington, DC 20590. If space provided for any item is inadequate, complete that item under Section IX, keying to the entry number being completed. Copies of this form, in limited quantities, may be obtained from the Information Systems Manager, Office of Hazardous Materials Transportation. Additional copies in this prescribed format may be reproduced and used, if on the same size and kind of paper.

**I. MODE, DATE, AND LOCATION OF INCIDENT**

1. MODE OF TRANSPORTATION: ❑ AIR ❑ HIGHWAY ❑ RAIL ❑ WATER ❑ OTHER

2. DATE AND TIME OF INCIDENT DATE: ______ TIME: ______

3. LOCATION OF INCIDENT
   CITY: ______ STATE: ______
   COUNTY: ______ ROUTE/STREET: ______

**II. DESCRIPTION OF CARRIER, COMPANY, OR INDIVIDUAL REPORTING**

| 4. FULL NAME: | 5. ADDRESS |
|---|---|

6. LIST YOUR OMC MOTOR CARRIER CENSUS NUMBER, REPORTING RAILROAD ALPHABETIC CODE, MERCHANT VESSEL NAME AND ID NUMBER OR OTHER REPORTING CODE OR NUMBER.

**III. SHIPMENT INFORMATION** (From Shipping Paper or Packaging)

| 7. SHIPPER NAME AND ADDRESS (Principal place of business) | 8. CONSIGNEE NAME AND ADDRESS (Principal place of business) |
|---|---|
| 9. ORIGIN ADDRESS (If different from Shipper address) | 10. DESTINATION ADDRESS (If different from Consignee address) |

11. SHIPPING PAPER/WAYBILL IDENTIFICATION NO.

**IV. HAZARDOUS MATERIAL(S) SPILLED** (NOTE: REFERENCE 49 CFR SECTION 172.101.)

| 12. PROPER SHIPPING NAME | 13. CHEMICAL/TRADE NAME | 14. HAZARD CLASS | 15. IDENTIFICATION NUMBER (e.g., UN 2764, NA 2020) |
|---|---|---|---|

16. IS MATERIAL A HAZARDOUS SUBSTANCE? ❑ YES ❑ NO
17. WAS THE RQ MET? ❑ YES ❑ NO

**V. CONSEQUENCES OF INCIDENT, DUE TO THE HAZARDOUS MATERIAL**

| 18. ESTIMATED QUANTITY HAZARDOUS MATERIAL RELEASED (Include units of measurement) | 19. FATALITIES | 20. HOSPITALIZED INJURIES | 21. NON-HOSPITALIZED INJURIES |
|---|---|---|---|
| 22. NUMBER OF PEOPLE EVACUATED | | | |

23. ESTIMATED DOLLAR AMOUNT OF LOSS AND/OR PROPERTY DAMAGE, INCLUDING COST OF DECONTAMINATION OR CLEANUP (Round off in dollars)

| A. PRODUCT LOSS | B. CARRIER DAMAGE | C. PUBLIC/PRIVATE PROPERTY DAMAGE | D. DECONTAMINATION/CLEANUP | E. OTHER |
|---|---|---|---|---|

24. CONSEQUENCES ASSOCIATED WITH THE INCIDENT: ❑ VAPOR (GAS) DISPERSION ❑ MATERIAL ENTERED WATERWAY/SEWER
❑ SPILLAGE ❑ FIRE ❑ EXPLOSION ❑ ENVIRONMENTAL DAMAGE ❑ NONE ❑ OTHER:

**VI. TRANSPORTATION ENVIRONMENT**

25. INDICATE TYPE(S) OF VEHICLE(S) INVOLVED: ❑ CARGO TANK ❑ VAN TRUCK/TRAILER ❑ FLATBED TRUCK/TRAILER
❑ TANK CAR ❑ RAIL CAR ❑ TOFC/COFC ❑ AIRCRAFT ❑ BARGE ❑ SHIP ❑ OTHER:

26. TRANSPORTATION PHASE DURING WHICH INCIDENT OCCURRED OR WAS DISCOVERED:
❑ EN ROUTE BETWEEN ORIGIN/DESTINATION ❑ LOADING ❑ UNLOADING ❑ TEMPORARY STORGE/TERMINAL

27. LAND USE AT INCIDENT SITE: ❑ INDUSTRIAL ❑ COMMERCIAL ❑ RESIDENTIAL ❑ AGRICULTURAL ❑ UNDEVELOPED

28. COMMUNITY TYPE AT SITE: ❑ URBAN ❑ SUBURBAN ❑ RURAL

29. WAS THE SPILL THE RESULT OF A VEHICLE ACCIDENT/DERAILMENT? ❑ YES ❑ NO

| A. ESTIMATED SPEED: | B. HIGHWAY TYPE:<br>❑ DIVIDED/LIMITED ACCESS<br>❑ UNDIVIDED | C. TOTAL NUMBER OF LANES:<br>❑ ONE ❑ THREE<br>❑ TWO ❑ FOUR OR MORE | **SPACE FOR DOT USE ONLY** |
|---|---|---|---|

Figure 5-2: DOT Hazardous Material Incident Report

**VII. PACKAGING INFORMATION:** If the package is overpacked (consists of several packages, e.g., glass jars within a fiberboard box), begin with Column A for information on the innermost package.

| ITEM | | A | B | C |
|---|---|---|---|---|
| 30. TYPE OF PACKAGING, INCLUDING INNER RECEPTACLES (e.g., Steel drum, tank can) | | | | |
| 31. CAPACITY OR WEIGHT PER UNIT PACKAGE (e.g., 55 gal., 65 lb.) | | | | |
| 32. NUMBER OF PACKAGES OF SAME TYPE WHICH FAILED IN IDENTICAL MANNER | | | | |
| 33. NUMBER OF PACKAGES OF SAME TYPE IN SHIPMENT | | | | |
| 34. PACKAGE SPECIFICATION IDENTIFICATION (e.g., DOT 17E, DOT 105A100, UN 1A1 or none) | | | | |
| 35. ANY OTHER PACKAGING MARKINGS (e.g., STC, 8/16–55–88, Y1.4/150/87) | | | | |
| 36. NAME AND ADDRESS, SYMBOL OR REGISTRATION NUMBER OF PACKAGING MANUFACTURER | | | | |
| 37. SERIAL NUMBER OF CYLINDERS, PORTABLE TANKS, CARGO TANKS, TANK CARS | | | | |
| 38. TYPE OF LABELING OR PLACARDING APPLIED | | | | |
| 39. IF RECONDITIONED OR REQUALIFIED | A. REGISTRATION NUMBER OR SYMBOL | | | |
| | B. DATE OF LAST TEST OR INSPECTION | | | |
| 40. EXEMPTION/APPROVAL/COMPETENT AUTHORITY NUMBER, IF APPLICABLE (e.g., DOT E1012) | | | | |

**VIIII. DESCRIPTION OF PACKAGING FAILURE:** Check all applicable boxes for the package(s) identified above.

41. ACTION CONTRIBUTING TO PACKAGING FAILURE

| | A | B | C | |
|---|---|---|---|---|
| a. | ❑ | ❑ | ❑ | TRANSPORT VEHICLE COLLISION |
| b. | ❑ | ❑ | ❑ | TRANSPORT VEHICLE OVERTURN |
| c. | ❑ | ❑ | ❑ | OVERLOADING/OVERFILLING |
| d. | ❑ | ❑ | ❑ | LOOSE FITTINGS, VALVES |
| e. | ❑ | ❑ | ❑ | DEFECTIVE FITTINGS, VALVES |
| f. | ❑ | ❑ | ❑ | DROPPED |
| g. | ❑ | ❑ | ❑ | STRUCK/RAMMED |
| h. | ❑ | ❑ | ❑ | IMPROPER LOADING |
| i. | ❑ | ❑ | ❑ | IMPROPER BLOCKING |
| j. | ❑ | ❑ | ❑ | CORROSION |
| k. | ❑ | ❑ | ❑ | METAL FATIGUE |
| l. | ❑ | ❑ | ❑ | FRICTION/RUBBING |
| m. | ❑ | ❑ | ❑ | FIRE/HEAT |
| n. | ❑ | ❑ | ❑ | FREEZING |
| o. | ❑ | ❑ | ❑ | VENTING |
| p. | ❑ | ❑ | ❑ | VANDALISM |
| q. | ❑ | ❑ | ❑ | INCOMPATIBLE MATERIALS |
| r. | ❑ | ❑ | ❑ | OTHER |

42. OBJECT CAUSING FAILURE

| | A | B | C | |
|---|---|---|---|---|
| a. | ❑ | ❑ | ❑ | OTHER FREIGHT |
| b. | ❑ | ❑ | ❑ | FORKLIFT |
| c. | ❑ | ❑ | ❑ | NAIL/PROTRUSION |
| d. | ❑ | ❑ | ❑ | OTHER TRANSPORT VEHICLE |
| e. | ❑ | ❑ | ❑ | WATER/OTHER LIQUID |
| f. | ❑ | ❑ | ❑ | GROUND/FLOOR/ROADWAY |
| g. | ❑ | ❑ | ❑ | ROADSIDE OBSTACLE |
| h. | ❑ | ❑ | ❑ | NONE |
| i. | ❑ | ❑ | ❑ | OTHER |

43. HOW PACKAGE(S) FAILED

| | A | B | C | |
|---|---|---|---|---|
| a. | ❑ | ❑ | ❑ | PUNCTURED |
| b. | ❑ | ❑ | ❑ | CRACKED |
| c. | ❑ | ❑ | ❑ | BURST/INTERNAL PRESSURE |
| d. | ❑ | ❑ | ❑ | RIPPED |
| e. | ❑ | ❑ | ❑ | CRUSHED |
| f. | ❑ | ❑ | ❑ | RUBBED/ABRADED |
| g. | ❑ | ❑ | ❑ | RUPTURED |
| h. | ❑ | ❑ | ❑ | OTHER |

44. PACKAGE AREA THAT FAILED

| | A | B | C | |
|---|---|---|---|---|
| a. | ❑ | ❑ | ❑ | END. FORWARD |
| b. | ❑ | ❑ | ❑ | END. REAR |
| c. | ❑ | ❑ | ❑ | SIDE. RIGHT |
| d. | ❑ | ❑ | ❑ | SIDE. LEFT |
| e. | ❑ | ❑ | ❑ | TOP |
| f. | ❑ | ❑ | ❑ | BOTTOM |
| g. | ❑ | ❑ | ❑ | CENTER |
| h. | ❑ | ❑ | ❑ | OTHER |

45. WHAT FAILED ON PACKAGE(S)

| | A | B | C | |
|---|---|---|---|---|
| a. | ❑ | ❑ | ❑ | BASIC PACKAGE MATERIAL |
| b. | ❑ | ❑ | ❑ | FITTING/VALVE |
| c. | ❑ | ❑ | ❑ | CLOSURE |
| d. | ❑ | ❑ | ❑ | CHIME |
| e. | ❑ | ❑ | ❑ | WELD/SEAM |
| f. | ❑ | ❑ | ❑ | HOSE/PIPING |
| g. | ❑ | ❑ | ❑ | INNER LINER |
| h. | ❑ | ❑ | ❑ | OTHER |

**IX. DESCRIPTION OF EVENTS:** Describe the sequence of events that led to incident, action taken at time discovered, and action taken to prevent future incidents. Include recommendations to improve packaging, handling, or transportation of hazardous materials. Photographs and diagrams should be submitted when necessary for clarificaiton. ATTACH A COPY OF THE HAZARDOUS WASTE MANIFEST FOR INCIDENTS INVOLVING HAZARDOUS WASTE. Continue on additional sheets if necessary.

| 46. NAME OF PERSON RESPONSIBLE FOR PREPARING REPORT | 47. SIGNATURE | |
|---|---|---|
| 48. TITLE OF PERSON RESPONSIBLE FOR PREPARING REPORT | 49. TELEPHONE NUMBER (Area Code) | 50. DATE REPORT SIGNED |

Figure 5-2 Continued

2. Participate in comparing classmates' forms and discussing reasons for discrepancies.
3. Participate with the class in selecting the most accurate form and brainstorming strategies for improving future reporting.

## Activity B

In class, place all of the DOT Hazardous Materials Incident Report forms completed in Activity A on a table. Next, students score each classmate's form for legibility and neatness using the following point system:

| | |
|---|---|
| 5 points | Perfect printing |
| 4 points | Not-so-perfect printing, but completely readable |
| 3 points | Printing looks suspiciously like cursive writing |
| 2 points | Hard to read |
| 1 point | Could not make out some words |

After each student's points are added, the average score is then calculated. Those achieving an average score of less than four should set a goal of improving this skill.

## Activity C

Prior to completing this activity, read Scenario 2 in Appendix 1.

Pretend that you are an EPIC consultant and write a brief memo to the UDL managers about the importance of their signatures on forms. As part of your memo, include a comparison of the signature block on the DOT Hazardous Materials Incident Report form (Figure 5-2) with the signature block from the EPA Notification of Regulated Waste Activity form (Figure 5-1)

# 5-2 Obtaining and Submitting Forms

## Concepts

- It is important to verify that a form is current so government agencies will receive the correct information.
- Using the Internet can expedite obtaining, completing, and submitting forms.
- After a form has been completed copies must be made, filed, and distributed.
- Information about the form should be entered in a tracking system.

Forms can be obtained and submitted either through the mail or the Internet. Regardless which method you choose, always be sure that you are completing the most current forms, which can be obtained by visiting a U.S. Government Bookstore, calling or writing the agency directly, or looking up the agency's Internet address and downloading the forms for printing.

Figure 5-3 provides a sampling of the current locations for some of the more common regulatory forms. Unfortunately, it is not unusual for the Internet address of an information source to move. Should this happen, or you cannot find a particular form, use one of the browsers, such as Yahoo, Excite, Infoseek, Lycos. etc. If you are sure of the form's correct title, then run a search using that name surrounded by double quotes (" "). If not, then run a search for the agency itself. At the agency's Web site look for a link to information about obtaining forms.

| Form Name | Form Number | Oversight Agency | CFR Reference (CFR.Part.Section) | Internet Location of Form (URL) |
|---|---|---|---|---|
| Hazardous Materials Incident Report | DOT Form F 5800.1 | DOT | 49.171.16 | http://hazmat.dot.gov/5800.pdf |
| Discard Monitoring Report | EPA Form 3320-1 | EPA | 40.122.21 | http://www.epa.gov/earth1r6/6en/w/dmrf.pdf |
| Notice of Intent (NOI) for Storm Water Discharges Assoicated with Industrial Activity Under an NPDES Permit | EPA Form 3510-6 | EPA | 40.122.26 | http://www.epa.gov/earth1r6/6en/w/dmrf.pdf |
| Underground Storage Tank Notice | EPA Form 7530 | EPA | 40.280.220 | http://www.epa.gov/swerust1/fedlaws/form7530.pdf |
| Pre-manufacturing Notice (PMN) | EPA Form 7710-25 | EPA | 40.720.40 | http://www.epa.gov/opptintr/newschms/pnmpart1.pdf (and) http://www.epa.gov/opptintr/newschms/pnmpart2.pdf |
| Pesticide Registration | EPA Form 8570-1 | EPA | 40.152.50 | http://www.epa.gov/opprd001/forms/8570-1.pdf |
| Pesticide Data Certification | EPA Form 8570-29 | EPA | 40.158.32 | http://www.epa.gov/opprd001/forms/8570-29.pdf |
| Notification of Regulated Waste Activity | EPA Form 8700-12 | EPA | 40.261 | http://www.epa.gov/epaoswer/hazwaste/data/form8700/notiform.pdf |
| Uniform Hazardous Waste Manifest | EPA Form 8700-22 | EPA | 40.262.20 | http://www.epa.go:80/epaoswer/hazwaste/gener/manifest/PDF/form.pdf |
| Hazardous Waste Permit Application | EPA Form 8700-23 | EPA | 40.270.10 | http://www.epa.gov/epaoswer/hazwaste/data/form8700.forms.htm |

Figure 5-3: URL Locations of Selected Governmental Forms

| Form Name | Form Number | Oversight Agency | CFR Reference (CFR.Part.Section) | Internet Location of Form (URL) |
|---|---|---|---|---|
| Toxic Chemical Release Inventory, Reporting Form R | EPA Form 9350-1 | EPA | 40.372.85 | http://www.epa.gov/opptintr/tri/toxdoc.pdf |
| Toxic Chemical Release Inventory, Form A (Emergency Planning and Notification, Tier I/Tier II/TRI) | EPA Form 9350-2 | EPA | 40.372.27 | http://www.epa.gov/opptintr/tri/toxdoc.pdf |
| Federal Operating Permit Application – General Information and Summary | Form GIS | EPA | 40.71.5 | http://www.epa.gov/oar/oaqps/permits/geninfo.pdf |
| Emissions Unit Description for Fuel Combustion | Form EUD-1 | EPA | 40.71.5 | http://www.epa.gov/oar/oaqps/permits/unitdesc.pdf |
| Emissions Unit Description for VOC Emitting Sources | Form EUD-2 | EPA | 40.71.5 | http://www.epa.gov/oar/oapqs/permits/unitdesc.pdf |
| Emissions Unit Description for Process Sources | Form EUD-3 | EPA | 40.71.5 | http://www.epa.gov/oar/oaqps/permits/unitdesc.pdf |
| Certification of Truth, Accuracy, and Completeness | Form CTAC | EPA | 40.71.5 | http://www.epa.gov/oar/oaqps/permits/comply.pdf |
| Material Safety Data Sheet | OSHA Form 174 | DOL/OSHA | 29.1910.1200 | http://www.osha-slc.gov/Publications/osha174.pdf |
| Log and Summary of Occupational Injuries and Illnesses | OSHA No. 200 | DOL/OSHA | 29.1910.1200 | http://www.osha-slc.gov/html/Forms/osha200/pdf |

Figure 5-3 Continued

In addition to the forms themselves, you will also need instructions for completing them. These can be obtained through the agency directly or on the agency's Internet site.

After you have found the correct form and completed it, ask the following important questions before putting it in the mail:

—Has the proper individual(s) signed the form(s)?

—Is the signed form(s) being sent to the appropriate agency?

—Have all of the instructions provided either on the form or in the form's instruction manual been followed?

—Is the form(s) being sent via certified mail with a return receipt, so you will have documentation verifying that the form(s) was received on a specific date?

—Have copied been made of the completed form(s)?

—Have copies of the form(s) been sent to the appropriate in-house managers or departments?

—Have copies of the form(s) been filed?

—Has data about the form(s) been input into a tracking system? Note that the tracking data should include information such as when the form was sent, who sent the form, what mail carrier was used, etc.

At this time only a handful federal agencies are experimenting with processes for submitting forms through the Internet. It is expected, however, that in the near future electronic transfer will be the standard method of submitting forms.

## Checking Your Understanding

### Activity A

Conduct a complete inventory of all potentially hazardous materials used and stored in your home (including living spaces, basement, and garage). If you prefer, you may perform this activity at your workplace, limiting the inventory to a single department if the company is large.

Find five chemical products suspected of being hazardous or labeled as hazardous and complete a

Chemical/Material Inventory Form (use either Figure 5-4 or Figure 5-5) for each one. Photocopy the form(s) as needed while conducting your inventory. Areas marked with an N/A (not applicable) on the form are not applicable to this activity.

| CHEMICAL/MATERIAL INVENTORY FORM |
|---|
| Generic Name: |
| Trade Name: |
| Supplier: |
| Address and Phone No.: |
| |
| |
| Label Statements: |
| |
| |
| MSDS Cross Reference No.: |
| MSDS received with shipment? ❑ Yes ❑ No<br>If no, attach copy of letter sent to manufacturer requesting MSDS. |
| Updated MSDS received? ❑ Yes ❑ No<br>If yes: (1) provide date of updated MSDS and date new MSDS supplies to MSDS file:<br>(2) does updated MSDS contain new health or physical hazard information? ❑ Yes ❑ No<br>If yes, provide date training is scheduled to explain new hazard information to workers: |
| Training needed? ❑ Yes ❑ No |
| Where used: |
| |
| Quantities: |
| |
| Where stored: |
| Date of inventory preparation: |
| Prepared/revised by (optional): |
| Annual review date: |

Figure 5-4: Sample 1 Chemical/Material Inventory Form

| CHEMICAL/MATERIAL INVENTORY FORM | | |
|---|---|---|
| Chemical Name: | | |
| Synonym(s): | | |
| Trade Name(s): | | |
| MSDS Reference: **N/A** | | |
| Supplier Name: | | |
| Address: | | |
| | | |
| Phone No.: **N/A** | | |
| **Information on Usage** | | |
| Location | Frequency | Quantity |
| | | |
| | | |
| | | |
| | | |
| **Storage Information** | | |
| Location | | Quantity |
| | | |
| | | |
| | | |
| | | |
| Date: | | |
| Review Dates: | | |
| | | |
| Preparer/Reviewer (optional): | | |

Figure 5-5: Sample 2 Chemical/Material Inventory Form

## Activity B

Use the Internet and CFR to complete Figure 5-6 and Figure 5-7. Since the information for the Internet portion of this activity is often available from different sources you may cite any URL that contains the actual form. The information for the CFR columns in the tables should be direct CFR section references.

| Form Name | Form Number | Oversight Agency | CFR Reference (CFR.Part.Section) | Internet Location of Form (URL) | No. of Pages in Form & Instructions (if applicable) |
|---|---|---|---|---|---|
| Discharge Monitoring Report | EPA Form 3320-1 | | | | |
| Underground Storage Tank Notice | EPA Form 7530 | | | | |
| Pre-manufacturing Notice (PMN) | EPA Form 7710-25 | | | | |
| Pesticide Data Certification | EPA Form 8570-29 | | | | |
| Hazardous Waste Permit Application | EPA Form 8700-23 | | | | |
| Toxic Chemical Release Inventory, Reporting Form R | EPA Form 9350-1 | | | | |
| Emissions Unit Description for Fuel Combustion Sources | Form EUD-1 | | | | |
| Certification of Truth, Accuracy, and Completeness | Form CTAC | | | | |
| Material Safety Data Sheet | OSHA Form 174 | | | | |

Figure 5-6: Activity Chart Containing the Form's Official Number

| Form Name | Form Number | Oversight Agency | CFR Reference (CFR.Part.Section) | Internet Location of Form (URL) | No. of Pages in Form & Instructions (if applicable) |
|---|---|---|---|---|---|
| Notice of Intent (NOI) for Storm Water Discharges Associated with Industrial Activity Under an NPDES Permit | | | | | |
| Pesticide Registration | | | | | |
| Notification of Regulated Waste Activity | | | | | |
| Uniform Hazardous Waste Manifest | | | | | |
| Toxic Chemical Release Inventory, Reporting Form R | | | | | |
| Federal Operating Permit Application – General Information and Summary | | | | | |
| Emissions Unit Description for VOC Emitting Sources | | | | | |
| Emissions Unit Description for Process Sources | | | | | |
| Log and Summary of Occupational Injuries and Illnesses | | | | | |

Figure 5-7: Activity Chart without the form's official number

## Summary

Companies dealing with hazardous substances are required by law to complete various forms. An inaccurate or incomplete form could jeopardize someone's health or life and cost employers a significant amount of money. Therefore, the ability to complete forms accurately and legibly is a critical skill to employers.

Forms are constantly being updated, so environmental technicians must ensure that they are completing the most current form. The Internet is a useful way to obtain forms, as well as transmit them quickly and efficiently. Most agencies have Web sites with locations where forms can be downloaded.

Many government forms require a certified signature. This signature should be from a professional engineer, an owner-operator, or an upper-level manager.

In addition to being completed and mailed, forms must be copied, filed, and tracked.

PRESERVING THE LEGACY

6

# Oral Communication

## Chapter Objectives

Upon completing this chapter, the student will be able to:

1. **Identify** three different types of oral communication within a business setting.
2. **Compare and contrast** three different oral presentation delivery methods.
3. **Plan and deliver** an effective formal presentation to an audience.
4. **Plan and deliver** an effective formal presentation to another person.
5. **Explain** the role of professionalism in oral communication.
6. **Understand** what harassment is as described in Title VII of the Civil Rights Act of 1964.

## Chapter Sections

# 6-1 Types of Oral Communication

## Concepts

- Oral communication skills are valuable to employers.
- In a business environment, the three basic types of oral communication are formal presentations to an audience, formal presentations to one person, and informal, unplanned communication.

In Chapter 1, you learned that the ability to communicate orally is a primary skill that employers value. While employers expect to provide some training, usually for implementing company policy regarding answering the telephone or responding to customer service calls, they prefer hiring individuals who already have well-developed oral communication skills. Companies seldom provide oral communication training for delivering formal presentations or conversing with fellow employees.

An environmental technician is likely to perform the following types of oral communication:

—A planned formal presentation to an audience

—A planned formal presentation to one person

—Unplanned, informal communication between one or more people

All formal presentations must be well planned and practiced. One type of formal presentation to an audience that an environmental technician makes is employee training. The planning for such a presentation involves the following:

—Analyzing the audience

—Determining the purpose of the training

—Creating a presentation outline

—Designing visual aids

A formal presentation to one person usually involves a supervisor or a client. For either of these presentations to be effective, the speaker should consider the purpose of the speech as well as the needs and abilities of the audience.

Unplanned, informal communications are those between colleagues and business associates. Liability or confidentiality problems occur when employees speak without carefully choosing their words or without the support of data. Informal communication can also convey verbal or nonverbal messages legally considered as harassment.

## Box 6-1 ■ Speaking Skills Are a Strong Asset in Every Field

**by Bill Quinnan**

We've all seen them – people that can walk into a room where they don't know a soul, and in minutes they are chatting with complete strangers like old friends.

Personally, I've always hated these extroverts. Or at least I would, if they weren't so darn likeable. The fact is, these people are at a distinct advantage in their careers. In their daily interactions with others, extroverts make a habit of impressing everyone around them – without bragging. By answering questions clearly, they satisfy those around them with thorough information. By speaking with confidence, extroverts leave others convinced that they are on top of things. Projecting that image can lead to promotions and new job opportunities as one consistently makes positive impressions on others.

Browse through the Smart Employment ads and you will see openings in fields as diverse as accounting, engineering, management and human resources, all calling for excellent verbal communication.

In fact, the Occupational Research Unit of the California Employment Development Department surveyed employers from 48 of California's top 50 occupations. Eighty-six percent of employers surveyed identified verbal communication as important. Whether employers were looking for janitors or teachers, accountants or mechanics, they felt that good speaking skills were necessary for the employee's success.

The reason is fairly obvious – an employee who speaks well can communicate observations, problems and ideas to the supervisor. In a customer-service environment, a good speaker can answer questions and help ensure that a customer gets the necessary information. A salesperson who speaks confidently about a product can pass on that confidence to the potential buyer.

The good news, for those of us who are more "conversationally challenged" is that good speaking skills can be learned. The ability to speak well can be broken down into smaller sub-skills. They include:

### Organizing Your Thoughts

Whether you are given time to prepare or are asked to answer a question on the spot, it is important to take a moment to gather your thoughts before you speak. This doesn't mean being silent for several minutes every time you are asked a question, but it means considering what information is relevant to the question at hand. If you are asked at an interview what skill was most important to the last job, think about examples that demonstrate when that skill was used. Conclude your answer by talking about how that skill might be useful at the job you are seeking. Your answer will be like a short speech, featuring a beginning, supporting information and an ending that satisfies the interviewer's curiosity.

### Knowing Your Subject Matter

In a job interview, this means learning as much as you can about the job and the company beforehand – and having a clear understanding of how your personal traits will serve the employer. This will help you tremendously when organizing your thoughts.

### Speaking with Clarity

A good voice begins at the diaphragm, not at the nose or even the throat. Practice filling your lungs with air before you begin speaking. Speak at a volume that is not boisterous, but makes it easy for others to hear you. Also, make an effort not to slur your speech or take shortcuts in pronouncing words. Don't say "I'm gonna;" say "I'm going to" or "I am going to." Also, try to avoid saying "uh" or "um" when speaking. It is better to say nothing until you know what you want to say.

### Using good grammar

Fair or unfair, people will often judge you by the words you use. If you make obvious errors like using double negatives ("I don't have none") or using "me" when you should use "I," you may turn off your listener. On the other hand, don't become overly obsessed with perfect grammar. A grammar book may tell you to say "I shall" instead of "I will," but you might sound like you are trapped in the wrong century if you do.

### Utilizing body language

Your gestures can help to communicate your ideas and command the attention of others. Your body movements should be smooth and natural; planning what body movements you will use and when you will use them probably will not work. Good body language is, for some, the hardest speaking skill to learn, because there are no solid rules about what gestures to use and when to use them. This is best learned by observing the gestures people use when speaking in everyday conversation, and trying to use effective gestures in your daily interactions. Don't forget to consider your facial expression when working on body language. Who would you rather talk to, a person with a warm, welcoming smile, or someone who wears a deadpan expression the entire time? If smiling doesn't come naturally to you, practice smiling in your daily interactions.

### Focusing on your subject

When asked a question, focus your attention on the task at hand: providing an answer. The more focused you are on this, the less attention you will be able to pay to your nervousness. If you ignore your jitters long enough, they will go away. If you are interested in improving your speaking abilities, consider joining a public speaking organization like Toastmasters International. While your goal may not be to get up in front of a room and give a speech, the practice you will get in Toastmasters will lend itself to better speaking habits. Regular evaluations of your speaking will help you improve your skills, including your ability to answer questions spontaneously. The number of Toastmasters chapters in Orange County alone are too numerous to list here, but they are listed on Toastmasters International's Web site – http://www.toastmasters.org – or you can call their headquarters (949) 859-8255.

# Checking Your Understanding

## Activity A

For the next week, observe a variety of people engaged in oral communication. Write a few paragraphs about the effects a setting has on each of the following:

—Tone of voice

—Choice of words

—Nonverbal actions

## Activity B

Analyze the tone, words, expressions, gestures (waving of hands, etc.), and body language used in your oral communications. Write a paragraph identifying a characteristic of your oral communications that needs to be improved.

# 6-2 Selecting a Delivery Method

## Concepts

- Formal presentations can be read from a script, memorized, or extemporaneous.
- Enthusiasm, expressed through voice inflections, keeps audiences interested.
- The best way to improve the delivery of a presentation is to practice, obtain feedback, and make corrections.

One of the first steps in planning a formal presentation is selecting a delivery method. The choices are as follows:

—Reading from a manuscript

—Memorizing a presentation

—Speaking extemporaneously

**Reading from a manuscript** is an accepted way to present papers that are expressly written for an organization or club. They are either composed of carefully crafted sentences or are technical in nature and intended to educate a specific audience. Papers prepared for literary or technical purposes are written to be heard so the words and ideas will be correctly stated and fully captured by the audience.

**Memorizing a presentation** takes a great deal of effort for most people, and the risk of forgetting material is a constant threat. However, if you are one of the few people who can easily memorize a large amount of material, then you may find this to be an effective method of delivering a speech. One advantage of memorization is that it frees you from notes, so you can maintain constant eye contact and vocal effectiveness. Another benefit is that it allows the exact delivery of a polished speech.

If you have decided that this is the best delivery choice, try the following memorization technique:

1. Read the manuscript several times, paying special attention to your vocal quality.
2. Tape record a practice reading, and play it over and over to aid in the memorization process.
3. Give a practice presentation to a family member or friend. Give him the manuscript and ask him to prompt you as needed.

**For an extemporaneous presentation** the speaker prepares an outline instead of a full manuscript. During the presentation, the outline prompts the speaker as he relays the information to the audience in a conversational style. Some people prefer to have a detailed outline in front of them as they speak; others use only a brief topical outline. For information that is particularly technical or literary, the speaker may still choose to read or memorize certain segments.

The advantages of a well-prepared extemporaneous presentation include the following:

—It allows for more eye contact with audience.

—The forgetfulness associated with memorization is not a problem.

—It facilitates a better, more natural use of nonverbal communication, such as gestures, posture, movement, and facial expressions.

**Choosing a delivery method** is a matter of personal choice. It also depends on the type of speech being delivered. Regardless of the delivery method employed, the key to bringing material alive is the effective use of eye contact and voice inflection. When speaking, make eye contact with the audience by talking to a group of people in the audience briefly and then moving on to another group. Do not focus on one person or object. Always look at and speak to the audience as if you were having a conversation with them. Do not let props get between you and the audience. For example, don't hide behind a podium during a presentation. If you are reading from a manuscript or notes, avoid placing them too low so you are constantly look down.

The best way to keep the audience listening is to interject enthusiasm into your speech. This is achieved through the inflections of your voice, which should be lively, but not overdone or exaggerated. The speaker who has a goal of sharing with the audience his personal enthusiasm and interest for a topic is likely to use the following vocal characteristics:

—Adequate volume (can be heard in the back of the room)

—Flexible pitch (avoids repetitious or sing-song patterns)

—Appropriate rate (does not read too fast)

—A pleasant tone (softens any harsh, grating, or shrill tendencies)

—Correct pronunciations (checks the dictionary for unfamiliar words)

—Clear articulation (tries not to slur, mush, or omit consonants)

Finally, practice is the most important element in delivering a speech. Do not think that because you are reading from a manuscript, have the words all memorized, or will have the notes in front of you that practice is unnecessary. It is required. Practicing allows you the time to make necessary corrections and to become comfortable with the material so your delivery will be more natural. Before giving a speech, do the following:

1. Read or practice the speech several times.
2. Make a tape recording of a session.
3. As you replay the tape, evaluate yourself on each of the voice characteristics mentioned above.
4. Correct obvious characteristics that detract from your presentation
5. Give the presentation to a friend and have him critique it.
6. Make corrections as necessary.

As you practice, clock your speech to be sure that it falls within the specified time allotted, as it is inconsiderate to speak over your time limit, especially when others are scheduled to follow.

## Checking Your Understanding

### Activity A

1. Create a short speech, or use a speech assigned by your instructor such as the *Gettysburg Address.*
2. Follow the six practice steps described in this section, and keep your first tape recording.
3. Record your final practice session.
4. Replay the first tape recording and the last tape recording.
5. Write a few paragraphs describing the differences.

### Activity B

Write a paragraph about the consequences of speaking beyond the time you have been allotted for your presentation. List the difficulties this can cause your audience, other speakers who will follow you, and the organization sponsoring the presentation.

# 6-3 Planning a Formal Presentation to an Audience

## Concepts

- Planning is the key to making effective presentations.
- Planning involves analysis of the audience and the purpose of the presentation.
- Presentations generally follow an outline that consists of an attention-getting introduction that states the presentation's purpose, a body containing the main points and supporting details, and a conclusion summarizing the main points and calling for action or a re-thinking of ideas.
- Visual aids are useful for organizing a presentation, keeping the audience's attention, and emphasizing important points.

Have you ever had to endure a presentation where the speaker spoke "off the cuff" and wandered aimlessly through the topic, or attended a presentation where the visual aids were so poorly prepared that they were only "visual" to those in the front row? Have you ever listened to a "speaker" who insisted on reading to you? What is unfortunate about all of these situations is that they could have been avoided with better planning.

In general, the more time you spend preparing a presentation, the more effective it will be (i.e., achieve its intended purpose). In addition to improving your presentation, thorough preparation can go a long way toward reducing those "butterflies" that interfere with an effective delivery. Ideally, preparation time should be spread over several days or weeks and involve the following:

—Analyzing the audience and purpose of the communication

—Preparing an outline

—Preparing the introduction, main body, and closing

To analyze the audience, try answering the following questions:

—How many people will be in the audience?

—What level of knowledge will the majority of the audience have about the topic?

—What presentation techniques can be used to make the topic more interesting to this particular group?

—How long will this audience be able to hold their attention to the presentation subject?

The size of the audience will determine if you should use the presentation technique of having each member of the audience introduce themselves and state their reason for attending. The knowledge of the audience about the subject matter will determine the level of the presentation. The fastest way to lose an audience is to either provide basic information they already know or talk over their heads. The appropriate level permits the listener to use existing knowledge as a springboard for understanding the new.

As you analyze your audience, think of various techniques for holding their attention. Some speakers can use humor effectively, others not. You could use graphics to gain or regain the attention of an audience of visual learners, or perhaps tell stories and anecdotes that are of interest to a particular group. For example, to describe to a group of gas station owners how even small leaks of hazardous substances can be expensive, you could tell the story of a gas station owner who controlled the weeds on his property with diesel fuel. After he poured about nine gallons of diesel fuel on top of the cracks in a concrete slab, it rained. A few days later several investigators from the Texas General Land Office arrived at the station to talk to the owner. The agency had detected diesel fuel in the town's water supply, and their investigators had traced the source back to the service station and fined him five thousand dollars.

Finally, when considering the audience's attention span, ask yourself if they will need or expect a break during the presentation. It is difficult and distracting for both the presenter and the audience to have people leaving and returning during a presentation. In general, if the presentation is going to exceed ninety minutes, then a scheduled break should be included. If the audience is highly motivated or well informed about the subject, you may get by with longer times between breaks. But remember that even for highly motivated audiences, there is a limit to the amount of information they can absorb.

To analyze the purpose of a presentation, ask the following questions:

—Why am I making the presentation?

—What actions do I want the audience to take?

—What information do I want the audience to receive?

—What can I do during the presentation to increase the probability that the audience will take the desired course of action?

The reason for making a presentation is usually to persuade or inform. If the reason is to persuade, then be clear about the message you want the audience to agree with or the action you want them to take. Often, deciding on the action that you want the audience to take will help determine the purpose of the message. For example, your informative presentation on recycling may change to a persuasive one if the action you want the audience to take is participating in a recycling program.

When deciding on what information you want the audience to hear, start by brainstorming a list of subjects, issues, anecdotes, etc. Then go over each item in the list, questioning its pros and cons and asking how useful it will be in delivering the message or making a topic more understandable. Next, combine information that is similar and distill the list into three to four main topics. While brainstorming for a persuasive presentation, think about ways in which you can tie the information to some action that you want the audience to take.

Finally, to increase the probability that the audience will take the desired action, try the following:

—**Additional Resources** Give the audience numbers to call or a bibliography for further information.

—**Marching Orders** Tell the audience the exact action that you want them to take.

—**Timing** Time the presentation so it is close to an event in which the audience can take action.

—**Useful Tools and Materials** Give the audience something that they are likely to use, such as a water sampling kit, a file folder filled with ample forms and information about how to fill them in, or special colored recycling bags.

—**Practice Desired Behavior** Have the audience perform the desired action while at the presentation. For example, in a problem solving presentation you could break the audience into small groups and have them role play the different techniques discussed.

—**Memorable Slogan** If a description of the action can be compressed into a few sentences or words, give the audience posters, buttons, magnets, or stickers with a slogan printed on them, such as the fire department's "Stop, drop, and roll."

—**Consequences** If your purpose is to get people to stop an unsafe behavior, you could tell them memorable, real-life stories. For example, if giving a presentation about electrical safety, talk about an industrial accident where an employee was careless or didn't follow the rules.

## Outline

The next step in preparing a presentation is creating an outline by using the format below:

I. Introduction
  A. Select an attention-getting opening, which could be a story, an interesting statistic, a fascinating fact, a thought-provoking question, or an audiovisual piece.
  B. State your thesis or primary purpose.
  C. Choose a means of relating the presentation to the audience.

II. Body
  A. State your first main point. Develop the point with details and audiovisuals, if applicable. Number the details (1., 2., etc.) below the main point as you write your outline.
  B. State your second main point and develop it.
  C. Continue with as many main points as needed (D., E., etc.).

III. Closing
  A. Summarize the main points.
  B. In a persuasive speech, relate the topic and then recommend an action to take or suggest a change of thinking to adopt.

**Thesis**

The thesis announces the primary reason for a paper or presentation. It tells the audience your position on a topic, which helps them follow your line of thinking.

Example of a thesis intended to inform:
"Our community's improved contingency plan has created a safer environment for citizens."

Example of a thesis intended to persuade:
"The U.S. Congress should take immediate and strong steps to address global warming."

## Introduction

Use the introduction of your presentation to state a purpose and provide a snapshot of where you plan to take the audience. Think of different ways to capture their attention. Is there an audio or visual aid you can use to illustrate the purpose of your presentation? For example, a park service employee could start a presentation on the reasons for protecting wetlands by playing a recording of all the birds that use them for their habitat. Species by species, their songs would be eliminated until there is only a single chirping bird and then absolute silence. After allowing the audience to sit uncomfortably in silence for a few seconds, the impact of loosing another habitat would be realized and the speaker could start the presentation.

Avoid starting any speech with, "I am going to talk to you about..."

Presentations are often introduced with an interesting quote or a piece of video. A story can serve the same purpose and in some cases be more effective. You can also use a bit of related humor or a joke to capture the audience. Use humor only if you have a good joke that is relevant to the topic and the ability to tell it effectively, as lame humor will do more to alienate your audience than to draw their attention. Select the story or joke very carefully making sure that there is a clear relationship between it and your topic. Avoid humor that is crude or could be considered a gender, racial, or religious slur.

## Main Body and Visual Aids

The body of the presentation is where the main points and detailed information to support them are located. Take time to fully research each point in the presentation. Collect information, examples, and statistics from reliable sources. Depending on the presentation, you may want to note sources to construct a bibliography.

If the topic you are discussing is controversial, then research all sides of the issue. As a speaker, you are not expected to be neutral. But, by presenting all sides of an issue and explaining why you favor one over the other, you will appear more credible and the audience will be inclined to accept your position as knowledgeable and well balanced.

Presentations should have some type of organization pattern so there is a logical flow of information from the introduction to the conclusion. The method selected should depend primarily on the audience and purpose. The following are some suggestions for possible sequencing orders:

—From simple to complex

—From general to specific

—From past to present.

Once an order for the presentation has been established, ask what stories or information would support the main points and help the audience relate to the topic. For example, a main point of a presentation could be that pollution prevention makes financial sense for small business owners. A way to support the point would be to provide statistics regarding what small businesses have saved by implementing a particular pollution prevention measure.

As you continue to develop the main body of your presentation, think about what types of visual aids you can use to accomplish the following:

—Support the main and supporting points

—Focus the audience's attention

—Help your audience remember key pieces of information

—Clarify complicated information

In addition, visual aids can be used to replace the traditional outline or note cards. In the past, public speaking classes often recommended that students place their main points and key information on 3 x 5 note cards. However, using notes tends to tie you to the podium and reduces your comfort to move about and interact with the audience. Outlines and note cards may also create unnatural breaks, because you must interrupt your flow to look down for the next point. A far better tactic is to place bulleted phrases on overhead transparencies or slides to trigger your thoughts. This enables you to relax and move around, adding energy to your presentation.

Recent technologies have improved the use of transparencies and slides. Software, such as Powerpoint™ or Presentations™, makes it easy to use visual aids and incorporate colorful backgrounds, different fonts, graphics, and limited animation. Wording can be made to fade-out, zoom-in, or pop-up as you talk about a point. Since the presentation is developed on a computer, it is easy to update or customize for a particular audience. The use of such strategies can not only make your presentation more interesting, but give it a more polished, high-tech appearance.

If you are inexperienced in the development of visual aids, here are some guidelines to consider:

—When preparing either transparencies or computer-projected slides, limit the amount of information and the number of lines of text on each projection. Consider the size of the projected image in determining the size of font to use. A good rule of thumb is to use fonts that are 14 points in size or larger.

—The background of the projected image should also contrast with the font (e.g., black text and white background). In larger rooms, what looks great on the monitor (e.g., yellow words against a blue background) does not necessarily project well.

—The background design should be kept simple, so it does not distract the audience from the text.

—If possible, perform a test run on the equipment in the room where you will give your presentation. Stand in different locations to determine the effectiveness of the graphic.

## Closing

The closing of your presentation should briefly summarize the body of the presentation by restating the purpose and reviewing the main points. If the presentation was designed to be informational, then the closing should highlight the future or explain the next course of action. If it was a persuasive presentation, then the closing typically urges the listeners to take action.

Inviting the audience to ask questions after the presentation provides an excellent opportunity for clarifying confusing points and obtaining audience feedback. Question and answer periods also have some disadvantages. For example, it can open your

### Box 6-2 ■ Tips for Using Projected Visual Aids

—Prior to the presentation, check the projector to ensure that it is properly focused.

—Position the projector so the images are easily visible from all parts of the room and the projector does not obstruct the view of the audience. Try placing the screen at a diagonal instead of directly in front of the projector.

—If you are left-handed, position the projector to your left, as you face the audience, or vice versa if you are right-handed. This allows you to place transparencies on the projector without turning your back on the audience.

—Face your audience, not the projected image on the screen. Always talk to the audience, not the wall.

—Be sure that a spare projector bulb is available.

—Tape the power cord to the floor to prevent members of the audience or you from tripping over the cord.

—Keep transparencies in a three-ring binder separated by a blank sheet of paper to protect them and keep them in order.

—Tell your audience what the next slide is before revealing it. This refocuses their attention as you move from one topic to another.

—Use a pointer to focus the audience on the next piece of information you want to highlight.

—If using transparencies, lay a piece of opaque paper over the transparency to conceal items you are not currently discussing. Slide the paper down to reveal each item you discuss.

## Box 6-3 ■ Repeat, Restate, and Reiterate

"Tell them what you are going to tell them. Tell them. Then tell them what you told them."

This advice has made the rounds among public speakers and writers. It serves to emphasize that the audience usually needs multiple exposures to remember and understand a point. This technique is referred to in public speaking and educational psychology books as repetition.

Repetition is also an effective tool for introducing new or complex ideas. One repetition strategy is to address the concept in the introduction, the body, and the closing of a presentation. Another is to discuss the concept in more than one way by using comparisons and analogies. The following is text from a fictitious presentation where the speaker is using repetition to teach the terms manufacturing input and manufacturing output. The repetition process consists of defining, comparing, and illustrating.

### Excerpt from a Presentation Using Repetition to Describe Manufacturing Inputs and Outputs

**Tell them the first time** by defining the terms.

... Inputs into a production process include energy and all the necessary materials, including those needed for equipment maintenance. Outputs include the product, unused energy, by-products, reusable or recyclable process materials, and wastes.

**Tell them a second time** by making a comparison or using an analogy.

We can compare what happens in our manufacturing process with how a steak is grilled on the barbecue. As I mentioned, the manufacturing inputs are all of the necessary materials including the material needed to maintain the equipment. So in my analogy "manufacturing inputs" would be the raw steak, seasonings, charcoal, starter fluid, matches, grill cleaner, scouring pads, and water.

I also mentioned that the "manufacturing outputs" include the product, unused energy, by-products, reusable or recyclable process materials, and wastes. So the cooked steak would be the product; the unused energy would be the heat rising from the barbecue; and the by-products would be the lighter fluid vapors and smoke. The reusable or recyclable process materials would be the half-used charcoal and matchsticks. The wastes would be the residue left on the grill, and the wastewater mixed with grill cleaner and scouring pad soap.

**Tell them the third time** by illustrating the manufacturing process with a flow chart showing the entry points of each input and point of generation for each output.

My next slide is a simple flow chart of an electroplating process used for the manufacture of chrome-plated faucets. At the top of the flow chart are the manufacturing inputs, the faucet castings, electroplating chemicals containing copper and chromium, water, and electricity. The center of the flow chart is similar to the barbecuing process. Here, in the first electroplating bath, electricity causes a thin layer of copper metal to form on the faucet's surface. After rinsing, a second electroplating bath adds a final layer of thin chromium metal.

The lower portion of the flow chart contains the manufacturing outputs. Instead of a cooked steak, the product is a chrome-plated water faucet. Instead of heat rising from the barbecue, the unused energy is the residual electrical energy from the electrolysis of water and heat from the electroplating baths. Instead of smoke, the by-product would be the water broken into hydrogen and oxygen gases from the electrolysis. Instead of half-used charcoal, the reusable or recyclable process materials would be clean water from counter-current rinse systems and some drag-out chemicals. Instead of wastewater mixed with grill cleaner and scouring pad soap, there would be hazardous chemical sludge at the bottom of each of the electroplating and rinse water baths.

presentation to an inquisition by a disgruntled or argumentative member of the audience. If that should occur, be sure to answer the questions as honestly as possible. If you don't know the answers, admit it. If the individual has a valid point, say so and don't be afraid to alter your point of view. Be prepared to cite your sources of information or provide additional facts. Do not loose your cool, and do not ridicule the individual. If the situation appears to be getting out of hand, offer to continue the discussion after the presentation and move to the next person with a question.

Try to keep within your allotted timeframe. If questions are still being asked at the end of your time, offer to continue the discussion later or encourage interested people to contact you by telephone or e-mail.

## Checking Your Understanding

### Activity A

Read Scenario 1 in Appendix 1.

As the Environmental Health and Safety technician for Locust®, you have been asked to generate a one-hour presentation on pollution prevention (P2) techniques for the company's office managers and clerical staff. Your audience will be composed of individuals who have no knowledge of P2, environmental issues, or the technologies and techniques that can be used to minimize or eliminate waste in an office. Develop the written outline for your presentation.

Consider the following questions as you developing the outline:

—What kind of terminology will be most appropriate for this audience?

—How and where can you use repetition or an analogy to teach the audience or drive home an important point?

—What types of examples will be appropriate for use in this presentation?

—What action do you want your audience to take after they leave?

—What are the anticipated difficulties your audience may encounter once they attempt that action?

—What resources can you offer to help the audience achieve the desired action?

Be sure to include an opening that grabs the attention of the audience. Within the outline, use parentheses to note ideas for the inclusion of visual aids as shown in the example outline in the next section.

### Activity B

Prepare an outline for a HAZCOM training presentation for an audience of ten workers who are experienced and familiar with manufacturing process terminology. Use the MSDS information provided for styrene (stabilized) found in Appendix 4. Include an introduction, main body, closing, and notes indicating what visual aids you would use and where you would make use of them.

# 6-4 Planning a Formal Presentation to One Person

## Concepts

- The planning for a formal presentation to one person is essentially the same as for a formal presentation delivered extemporaneously to an audience.
- Visual aids are useful for one-on-one presentations, but should not be elaborate.
- When planning a one-on-one presentation always keep to the allotted time, leaving enough time for questions.

Environmental technicians give one-on-one presentations more frequently than presentations to groups. One-on-one presentations include explaining a work practice to an inspector, reviewing a product with a company representative, providing customer service, and discussing a problem or salary with a supervisor. During these exchanges, the purpose of the communication can be to inform or persuade.

When selecting words and preparing for one-on-one presentations, think about the characteristics of the person to whom you will be speaking and the purpose of the communication. Try to communicate all of the information in a reasonably brief period of time. If under a time constraint, plan the presentation in such a way that there will still be time for questions and clarification.

Since one-on-one presentations are all extemporaneous, your planning will include the creation of an outline. In that outline, make use of the same elements (introduction, body, and close) that are used for audience presentations. The following is a model:

—The introduction of the presentation should get attention and relate the topic to the listener.

Example:

"I would like to tell you about a way to save the company money. It involves making a process change that will reduce the amount of hazardous waste we generate. The new equipment required for this change is a solvent distillation unit that can be amortized in only two years through the reduction in our hazardous waste disposal costs. After that, the savings will be added to our bottom line."

—The body of the presentation develops the main points, supported by details.

Example:

I. Explain the principle and practice of a solvent distillation unit.
   A. Recycled solvent is pure enough to be reused. (Show graphic on research data.)
   B. The company has an appropriate facility for the safe operation of the unit.
   C. The labor required to operate and maintain the distillation unit is comparable to the labor used to manage hazardous waste. (Show information that documents time spent on waste management.)

II. Illustrate current operating expenses associated with solvent use and disposal. (Show a table comparing all costs associated with solvent use and disposal.)
   A. Describe the material costs.
   B. Describe the waste disposal costs.

III. Illustrate a projection of one month's operating expenses using solvent distillation. (Show a table comparing all costs associated with solvent distillation.)
   A. Describe the material costs and compare it with the solvent use and disposal method.
   B. Describe the waste disposal costs for still bottoms (i.e., distillation waste, which is small in quantity) and how it is less than solvent use and disposal.

IV. Review literature from several equipment vendors. (Show a table illustrating A and B below.)
   A. Calculate the investment payback period.
   B. Calculate future savings after payback period.

Note: The graphics used to enhance the main points for one-on-one presentations do not have to be elaborate.

—The closing should summarize the main points and be designed to get the listener to take action.

Example:

"An additional benefit to starting a solvent recycling program is that it will change the EPA hazardous waste status of the company to a small quantity generator (SQG). Under the 1984 amendments to Resource Conservation Recovery Act (RCRA), if a business can reduce its hazardous waste generation to less than 1,000 kg per month, it is classified as a small quantity generator (SQG). Being an SQG exempts businesses from the full hazardous waste manifest provisions, allows them to store wastes on site for a longer time, and reduces the requirements for employee training, emergency planning, and contingency planning."

## Checking Your Understanding

### Activity A

Pretend that you are the technician who created the example outline in this section and that you have successfully convinced your boss to switch to solvent distillation. Prepare a presentation to your boss (i.e., the class instructor) for which the purpose is to get a raise. Outline the presentation; practice your delivery; and then make the presentation in class. (You may be creative in terms of the solvent recycling profits, the size of the company, your responsibilities, and your salary.) Afterward, listen as the class discusses the strengths and weaknesses of your content and delivery.

### Activity B

Read Scenario 1 in Appendix 1.

Study OSHA's Hazard Communication requirements in 29 CFR 1910.1200 in Appendix 2. Prepare an oral presentation for Mr. Crabgrass about a) the required elements in a HAZCOM program and b) some of the current practices at Locust® that need correcting.

# 6-5 Unplanned, Informal Communication

## Concepts

- When talking informally at work always think before you speak – consider your surroundings, what you are saying, and to whom you are speaking.
- Do not talk about confidential or potentially libelous information to people who work outside of the company. Always refer them to a company spokesperson or supervisor.
- Harassment is illegal.
- Know and follow company communication policies and the EEOC's rules regarding ethnic and sexual harassment.
- If in doubt about whether a comment would be considered harassment, then do not say it.
- If someone harasses you, then report the incident to a supervisor.

Employers and employees participate in all kinds of unplanned work-related communication. Such communication tends to be impromptu; that is, prompted by the occasion rather than planned. Although impromptu communications are a natural part of conversation, they can pose some unexpected dangers. On occasion human frailties can intrude. Lack of sensitivity can alienate workers. A tendency to gossip can result in unprofessional behavior. Not thinking about what you are saying or your surroundings can cause a slip of the tongue.

An actual example of such a situation involved a technician who was talking to a friend at lunch about the details of his work on a project site. He did not realize that the people in the next booth were employees from the Department of Natural Resources (DNR) and associated with the project. Because the information the technician was sharing in the conversation breached the client's confidentiality, the DNR representatives felt compelled to notify his boss. As a result, management developed and implemented a set of policies regarding client confidentially that require all new employees to receive training so they will understand the need for confidentiality.

Environmental technicians are often knowledgeable of information that is confidential or potentially libelous to their company. To handle sensitive communications, companies often designate a spokesperson to deal with outside agencies or the press, should the need arise. Spokespersons are often middle managers who have excellent oral and written skills, and the ability to remain calm in a crisis. They know their companies' basic processes and can be trusted to conduct themselves professionally. Therefore, all confidential or sensitive questions, particularly from sources outside the company, should be referred a spokesperson. In addition, it is recommended that all conversations regarding confidential subjects with outsiders be documented in field notes.

Imagine that a news reporter approaches when you are taking soil samples at a client's industrial site. The reporter asks if you expect to find anything dangerous to the public. If your company has properly prepared you, then your response would be, "May I suggest that you speak with Pat Jones. Pat has the background to address your questions." You could even provide the reporter with contact information. Such an approach removes you from a situation where you could divulge confidential information.

If you find yourself working as an employee or consultant for a company that does not have an established policy addressing confidential issues and an outsider approaches you, then respond as follows:

—Pause and think about what you are going to say

—Refer all confidential matters to your superior or client

—Document the interaction in your field notes.

## Avoiding Communication That Harasses

In recent years, the issues of discrimination and harassment in the workplace have received much attention. Title VII of the Civil Rights Act of 1964 addresses the rights and responsibilities of employees and employers in matters of discrimination and harassment. The federal government's Equal Employment Opportunity Commission (EEOC) is charged with educating the public and enforcing the regulations associated with the Civil Rights law. Most instances of workplace harassment are the product of a racial slur, vulgar sexual proposition, swearing, or other inappropriate communication. Regardless of whether they are employees or employers, environmental technicians must understand and follow laws concerning harassment, as they apply to everyone.

## Ethnic Harassment

According to the EEOC, harassment based on national origin, race and/or color violates Title VII. Ethnic slurs, racial jokes, offensive or derogatory comments, or other verbal or physical conduct based on an individual's race/color constitutes unlawful harassment if the conduct creates an intimidating, hostile, or offensive working environment, or interferes with the individual's work performance.

Equal Employment Opportunity Commission (EEOC) areas of regulation include the following:

— Sexual Harassment
— Race/Color Discrimination
— Age Discrimination
— National Origin Discrimination
— Pregnancy Discrimination
— Religious Discrimination
— Americans with Disabilities

Employers have a responsibility to maintain a workplace that is free of racial and national origin harassment. According to EEOC, employers may be responsible for any on-the-job harassment by their agents and supervisory employees, regardless of whether the acts were authorized or specifically forbidden by the employer. Under certain circumstances, an employer may be responsible for the acts of non-employees who harass their employees at work[1].

## Sexual Harassment

EEOC has the following to say about sexual harassment[2]:

"Sexual harassment is a form of sex discrimination that violates Title VII of the Civil Rights Act of 1964.

"Unwelcome sexual advances, requests for sexual favors, and other verbal or physical conduct of a sexual nature constitutes sexual harassment when submission to or rejection of this conduct explicitly or implicitly affects an individual's employment, unreasonably interferes with an individual's work performance or creates an intimidating, hostile or offensive work environment.

"Sexual harassment can occur in a variety of circumstances, including but not limited to the following:

—The victim as well as the harasser may be a woman or a man. The victim does not have to be of the opposite sex.

—The harasser can be the victim's supervisor, an agent of the employer, a supervisor in another area, a co-worker, or a non-employee.

—The victim does not have to be the person harassed but could be anyone affected by the offensive conduct.

—Unlawful sexual harassment may occur without economic injury to or discharge of the victim.

—The harasser's conduct must be unwelcome.

"It is helpful for the victim to directly inform the harasser that the conduct is unwelcome and must stop. The victim should use any employer complaint mechanism or grievance system available.

"When investigating allegations of sexual harassment, the EEOC looks at the circumstances, such as the nature of the sexual advances, and the setting in which the alleged incidents occurred. A determination on the allegations is made from the facts on a case-by-case basis.

"Prevention is the best tool to eliminate sexual harassment in the workplace. Employers are encouraged to take steps necessary to prevent sexual harassment from occurring. They should clearly communicate to employees that sexual harassment will

[1] http://www.eeoc.gov/facts/fs-race.html and /fs-nator.html, February 1999.

[2] http://www.eeoc.gov/facts/fs-sex.html, February 1999.

not be tolerated. They can do so by establishing an effective complaint or grievance process and taking immediate and appropriate action when an employee complains."

The University of Denver's Internet site (http://www.du.edu/eeoaa/harassment.html) offers clearly written advice regarding harassment and suggests the following:

"Simply speaking directly to the person who made the unwelcome overtures of offensive jokes or comments is often enough to put an end to the behavior that, if allowed to continue, creates a hostile or offensive environment. Direct confrontation gives you a degree of power, as well as responsibility, for controlling and avoiding situations that might degenerate into sexual harassment."

If communicating your reaction does not achieve the desired results, then refer to the University of Denver's Internet site above. It contains policies for making informal and formal complaints that may be useful.

Harassment is a topic that is not always handled well by employers and employees. Too often, discussions have been known to degenerate due to a lack of sensitivity, understanding, and experience. The atmosphere created by discussions that go astray can be just as painful to an employee as the harassment experience. A well-written, firm, and fair company policy distributed to each employee is often the best way for an employer to prevent harassment. Worthwhile examples of policies, such as the University of Denver's, are available on the Internet.

### When Verbal Harassment Threatens to Turn Violent

OSHA is giving attention these days to the threat of workplace violence. Environmental Safety and Health personnel may soon find themselves incorporating the topic into their training schedules. Although the subject is beyond the scope of this book, information resources regarding workplace harassment that threatens to turn violent are available. On the Internet, information can be found at http://www.osha-slc.gov/OshDoc/Other_Agency_data/C19941222.html, which is entitled "Update on Workplace Violence Labors Coalition Meeting with OSHA Officials." For titles of books, refer to this book's bibliography.

## Checking Your Understanding

### Activity A

Contact a company or environmental professional and ask if they have a written policy that addresses such things as press releases and maintaining client confidentiality. If possible, obtain a copy of the policy. Write a report summarizing your findings.

### Activity B

Contact an environmental company to determine if they have a spokesperson. If possible, interview a spokesperson about responsibilities and the type of training or experience required for their position. Prepare a three-minute oral presentation for the class. Submit the outline for your presentation to the instructor.

### Activity C

By either contacting a local company or conducting a Web search, obtain a model that could be used for developing a company's harassment policy. In a paragraph or two describe the key elements that must be included in such a policy.

## Summary

Oral communication on the job involves both planned and unplanned situations. At work, planned communication may include formal presentations to an audience and/or to one person, such as a customer, supervisor, or government inspector. For formal presentations, begin by analyzing the audience and the underlying purpose of speaking to the audience.

A formal presentation contains the following:

—An introduction designed to get the audience's attention, state the primary purpose, and relate personally to the audience.

—A body that contains several main points with supporting details.

—Visual aids, if applicable.

—A closing that summarizes the main points. If the presentation is persuasive in purpose, then the closing recommends an action or change of thinking that the audience might adopt.

You may select among several methods for delivering a planned presentation: reading a manuscript, memorizing the speech, or speaking extemporaneously. Each requires practice, the use of eye contact, and effective use of the voice. The effective voice is one that demonstrates interest in and enthusiasm for the material.

Most on-the-job oral communications are informal and unplanned rather than formal and planned. Sometimes human weaknesses can cause inappropriate remarks during impromptu communications. To decrease the chance of saying the wrong thing during an unplanned oral communication, prepare yourself by adopting habits of professionalism and by understanding company policy regarding unplanned oral communication. When you must communicate in a situation that may be associated with client confidentiality or with potential liability, refer individuals to a company spokesperson and document the conversation in your field notes.

Employers and employees should know what ethnic or sexual harassment is. The Equal Employment Opportunity Commission defines these types of harassment and makes recommendations for prevention and management of these situations. Companies should establish and provide employees well-written, firm, and fair company policies regarding the management of harassment.

PRESERVING THE LEGACY

7

# Communication Skills Overview

## Chapter Objectives

Upon completing this chapter, the student will be able to:

1. **Describe** current job search trends.
2. **Develop** a portfolio that contains a collection of work highlighting applicable skills and experiences.
3. **Develop** a résumé that is clearly written, neat in appearance, well organized, and emphasizes the positive aspects of your skills and employment history.
4. **Write** an effective cover letter.
5. **Explain** what steps need to be taken prior to participating in an interview.
6. **Respond** in a professional manner to commonly asked and inappropriate interview questions.

## Chapter Sections

# 7-1 Job Search Trends

## Concepts

- The basic steps in obtaining a job are creating a portfolio, writing a cover letter and résumé, and going to an interview.
- Résumés should only reflect job skills, experience, and training.
- Résumés should be no longer than two pages.
- Because of litigation, portfolios are often used instead of references.

Although the basic elements of the job application process (creating a portfolio, writing a cover letter and résumé, and going to an interview) have remained unchanged, trends continue to come and go. What is considered acceptable today may not be acceptable tomorrow. For example, in the past it was acceptable, and even beneficial, for applicants to note marital status on résumés since married people were considered more stable than those who were not. Today employers are more sensitive to laws involving discrimination, so it is now customary to omit all information regarding marital status as well as gender, age, and family (i.e., if you have dependents).

Since employers should evaluate applicants on skill, experience, and training alone, résumés should reflect that and only contain information pertaining to those areas.

A second trend is in the acceptable length of a résumé. In the past, few employers placed limits on what they would accept. During the period of downsizing and high unemployment in the 1990s, employers found themselves having to review hundreds of applications. Out of this glut grew the current trend to limit résumés to a maximum of two pages for most positions and no more than four pages for upper management positions.

A third trend is the increased use of portfolios. In the past, employers relied on references from previous employers. The growing fear of litigation has spawned a trend within human resource departments to limit discussion of an employee's work record to simply verifying the position held and beginning and ending dates of employment. Employers are now using the portfolio to fill the resulting information gap.

Prior to the start of any job search effort, it is wise to research current hiring trends by contacting your library, college placement office, or one of the job search sources on the Internet. In addition, many placement companies (headhunters) have Internet sites, which, in addition to recruiting applicants, offer articles on résumé trends and evaluation services. While the motivation behind their offerings is to expand their pool of employable applicants, these companies can still provide valuable information and assist in finding a job.

## Checking Your Understanding

### Activity A

Use the Internet, the library, and the school placement office to gather information about job search trends. Write a list of the current trends and share them with the class.

### Activity B

Using the Internet, the library, and the school placement office as resources, write a list of the types of information that should be included on a résumé and a list of information that should not.

### Activity C

The job placement service at your college can be an excellent resource for advice and information on career trends, potential employers, résumé and portfolio development, and interview skills. Make an appointment with the placement office to discuss environmental technician hiring trends. Be prepared to discuss your skills, experience, and training. Find out what services are provided by the placement office and have them help you determine what you need to do to improve your chances of getting a job. Prepare a written report summarizing your findings.

# 7-2 Creating Portfolios

## Concepts

- Portfolios allow employers to see solid evidence of an applicant's skills.
- A well organized, professional looking portfolio shows an employer the quality of your work, your ability to communicate, and that you can complete a large project.
- Portfolios can include examples of school work, projects, newspaper clippings, letters of recommendation, field notebooks, etc.
- Tailoring your portfolio to match specific job requirements tells an employer that you have researched the job and thought about how your experience applies to it.

"My first experience with a portfolio occurred in the early 1990s when I was hiring an administrative assistant. I had five or six qualified finalists to interview. The first two applicants assured me they could build a database, knew desktop publishing, etc. However, the third interviewee arrived with the usual résumé and transcript as well as a thick, three-ring binder.

"During the interview, Ms. Vincent asked if I was looking for any special skills. I told her yes, that part of the responsibilities of the position included developing promotional materials for the programs I administered and managing a collection of databases. She flipped through her portfolio and began showing me pamphlets and brochures she had developed for her previous employers. She then flipped to another section of the portfolio and showed me a database she had designed. She explained that the database could be merged into letters, to make mailing labels or name tags, and could be sorted by name, affiliation, zip code, etc.

"I was impressed – not only by the quality of work, but by the fact that she had thought to show the material as part of her interview. At the end of the interview, I had a real sense of Ms. Vincent's skill levels. The other applicants had told me what they could do, but Ms. Vincent showed what she was capable of doing. She got the job and, I'm happy to say, still works for me."

Lea Campbell
SE PETE Regional Director

One of the more effective job search tools is a portfolio. A portfolio is a collection of your work compiled over a period of time. Portfolios are important to employers because they show concrete evidence of your skills. A GPA and transcripts only give employers a vague idea of your level of expertise because course titles provide little information about what has been taught. But, a portfolio provides a window into what you can do – and that is what the employer needs to know.

In addition to demonstrating skills, a portfolio sends an important message to interviewers. It tells them that you have researched the job and thought about how your experience applies to the required skills. For employers, interviewing and hiring is a frustrating, time-consuming process with expensive consequences if the wrong person is hired. By providing a portfolio, you can make the decision easier and help assure them that you are best qualified for the position.

Employers also place a high value on communication and reporting skills. A portfolio is a good way to provide evidence of these difficult-to-document skills. The overall structure and appearance of the portfolio shows an employer your ability to collect, organize, and present data accumulated over a period of time. Portfolios also show that you can tackle a large, tedious project and follow it through to completion in a professional, organized manner.

To create an effective portfolio, collect materials from the beginning of your college or professional career. As you complete assignments, remember that while an immediate goal may be to learn the material and satisfy the requirements of the course, the ultimate goal is to create documents that will impress a potential employer. The following are suggested ways to build a portfolio:

—Type school reports and presentations, even if it is not required, so they may be used as examples later.

—Make sure field and lab notebooks are neatly written and maintained according to industry standards.

—Keep detailed notes or develop written reports describing any special projects that you complete for a class, internship, or job.

—Keep clippings of local newspaper or student publications that highlight any of your projects.

—Take photographs of projects that you have designed and completed.

Another good way to build a portfolio is to intern either formally for a class or informally by working for a local business or municipal agency. Then, document the internship with a report highlighting the skills that you have learned and pictures of you working at the host site. Also, try to obtain a letter of reference from your host.

Portfolios should include letters of recommendation. These letters can be written by former colleagues, who can comment on your competency; former bosses, who can tell about your contributions to the company; or an instructor who has had the opportunity to see the quality of your work. Letters of recommendation serve not only as a record of your education and employment, but a valuable tool for competing with other equally qualified candidates for a job.

Store portfolio materials in a box or file cabinet, so they won't be damaged or lost over time. A semester before beginning to look for a job, organize the portfolio by selecting pieces that best reflect your skill levels. The most effective organizational method is to order the materials by job skills such as sampling, emergency response, and Hazard Communication training.

Present the materials in a large three-ring binder that has a conservative plastic cover. Binders with clear, plastic pockets on the front are useful because you can slip a copy of your résumé in as the front cover. Prepare a tabbed divider page for each section. If you don't put your résumé on the front of the binder, then the first section should have it and a copy of your transcript. If you are organizing sections by skill, arrange them in alphabetical order so the interviewer can find information quickly. Mount any pictures, odd sized charts, or graphs on white bond paper and be sure each has a typed caption. At the end of the portfolio include photocopies of materials, such as letters of recommendation, to leave with the employer.

## Box 7-1 ■ Putting Portfolios on Disk

The availability of more powerful computer hardware, scanners, and graphic/photographic software is allowing job seekers the opportunity to turn traditional bound portfolios into interactive, electronic documents. Within these electronic portfolios, the introductory page is set up much like a home page on the Internet. On this page is a table of contents and topics that are linked by buttons to various parts of the portfolio. Therefore, if an employer were interested in looking at your sampling and analysis skills, he would click on that button and the portfolio would open to the section describing those skills.

The electronic portfolio has a number of advantages. It is an extremely impressive and professional way to present your skills. Electronic portfolios allow you to make a very colorful and eye-catching document into which photographs and video clips can be scanned. When employers view an electronic portfolio, they can see the value of hiring an employee with the skill and creativity to present in this format. Electronic portfolios can be e-mailed to a potential employer. It is easy and inexpensive to make multiple copies of the portfolio so that you can leave a copy with the interviewer for in-depth review at a later time.

On the down side, you may not have the skills to develop such a document. In that case, you might consider hiring a student from the business information systems or computer science department to prepare it for you. Most Internet service provider companies that develop and maintain home pages will also be able to develop an electronic portfolio. You can locate them in your telephone directory under "home page management" or "Internet service provider."

An electronic portfolio is only an asset if the employer has the hardware and software to open such a document. ZIP disks or CD ROMs allow you to include more pictures and compose longer electronic files, but don't send your information on one of these unless you are certain that the receiver has access to the proper computer drives. For those employers who may not have the hardware or software to view your document, it is wise to have an old-fashioned bound version.

As with any document you plan to present to a potential employer, neatness and organization are critical. Have an instructor or counselor review the finished portfolio for both appearance and content. Information regarding how to use a portfolio during the interview is included in the Interview section of this chapter.

## Checking Your Understanding

### Activity A

Create an outline for a portfolio. Start by listing all of the examples of your work, letters of recommendation, and photos that you would include in the portfolio. Next, organize the list into an outline using top level headings to indicate tabs.

### Activity B

Prepare a portfolio using the outline in Activity A. Keep in mind that the contents of the binder may continue to change, but not the basic organizational structure.

# 7-3 Writing Résumés and Cover Letters

## Concepts

- The purpose of cover letters and résumés is to obtain a job interview.
- There are two basic types of résumés: chronological and functional.
- Gaps in employment can be filled in with volunteer work, continuing education, and consulting.
- When writing a résumé that will be scanned into a database, enter as many key words used in the job description as possible.
- Since it is the first thing a prospective employer reads, cover letters are as important as a résumé.

A résumé is a brief history of your accomplishments that is designed to get an interview. To achieve this goal, it must be neat, clearly written, well organized, and emphasize the positive aspects of your skills and employment history. Résumés for entry-level positions should be limited to two pages.

Begin your résumé with a heading that gives your full name, address, telephone number, and e-mail address. Next, if you are applying for a specific position, is a job objective statement that summarizes why you are applying for it. A well-written job objective statement sets up the remainder of the résumé to describe how you are qualified to meet the job objective. An example of a good job objective is as follows:

> A technician position in the environmental division of a chemical manufacturing company.

Note that the example is specific about the job type, division level, and type of company for which the applicant would like to work.

Before building the body of the résumé, take the time to find out what skills and experience are required for the job by studying the classified advertisement and requesting a copy of the job description from a company directly. One of the best ways to learn about a position is to talk with someone already working in the field. Such efforts will help in customizing the résumé so the skills listed are the ones required for the job.

When thinking about the content of your résumé focus on how your education, experience, and training have prepared you for the position. To ensure that important information is not omitted, use the following checklist:

| ✓ | Résumé Content Checklist |
|---|---|
| | Names, dates, and locations of schools attended |
| | Area(s) of study (major and minor) |
| | Degrees, certificates, or licenses received |
| | Grade point average |
| | Names and addresses of employers |
| | Dates of employment |
| | Job and title of duties |
| | Major accomplishments |
| | Awards and special related skills |

## Résumé Formats

Currently, two basic résumé formats are in used: chronological and functional. The chronological style (see Figure 7-1) arranges jobs in the order in which they have been held, with the most recent listed first. The advantage of this format is it provides a clear picture of where you have worked and your accomplishments. It is also orderly, logical, and easy to follow. The disadvantage is gaps in employment history are easy to spot. Also, a chronological résumé does not highlight skill areas.

In a functional résumé (see Figure 7-2) work experiences are arranged according to skill areas. The advantage of this approach is that it draws attention to your skills. A disadvantage is that some employers look at functional résumés suspiciously because they hide employment gaps.

**Clyde Barrow**
**123 East Street**
**Willow Hills, Texas 77631**

**(409) 555-4321**

**Objective**
A training specialist in environmental/safety services.

**Education**
Willow Hills Community College, Willow Hills, Texas
AAS Environmental Safety Technology, May 1999
GPA: 4.00

**Experience**

(January 1999 – May 1999)
Kingston Waste Management
Intern
Responsibilities included soil and water sampling and maintaining sampling log. Also assisted with environmental/safety training.

(September 1998 – December 1998)
Willow Hills Small Business Development Association
Intern
Performed pollution prevention audits for small businesses.

(August 1997 – Present)
Industrial Training Association
Trainer
Responsible for managing ITA's computerized environmental/safety training lab.

(August 1997 – September 1998)
Willow Hills Community College
Tutor
Responsible for peer tutoring students in the areas of math, chemistry, and environmental technology.

**Special Skills**
Fluent in Spanish
Proficient with Microsoft Office 97 and Wastetracker 2.0 software
Current 40-hour HAZWOPER Certificate

Figure 7-1: Chronological Résumé Example

**Clyde Barrow**
**123 East Street**
**Willow Hills, Texas 77631**

**(409) 555-4321**

**OBJECTIVE:**
A training specialist in environmental/safety services.

**SKILLS AND ACCOMPLISHMENTS:**

- Delivery and Development of Training
  Managed and monitored computerized safety training lab. Responsible for scheduling, monitoring students, installing software, and minor PC maintenance. Assisted in providing eight-hour update training. Developed and delivered pollution prevention workshops for small business owners. Fluent in Spanish and have served as translator during eight-hour update for Spanish speaking employees.

- Sampling and Analysis
  Performed soil and water sampling. Maintained sampling tracking log. Responsible for labeling, organizing, and storing samples.

- Audits
  Performed pollution prevention audits for 30 small businesses.

- Computer Skills and Certificates
  Proficient with Microsoft Office 97 and Wastetracker 2.0.
  Current 40-hour HAZWOPER Certificate

**Work History**
Kingston Waste Management
Willow Hills Small Business Development Association
Industrial Training Association

**Education**
Willow Hills Community College, Willow Hills, Texas
AAS Environmental Safety Technology, May 1999
GPA: 4.00

Figure 7-2: Functional Résumé Example

If there is a gap in your employment record, try to recall if you were doing anything that could be presented as work. If so, create a reasonable job title, and include it. For example, if you were caring for children, then create an entry on your résumé such as the following:

1998 – 1998 Caregiver: Full-time care of three small children.

If currently unemployed, consider doing volunteer work related to your field, taking continuing education classes, or doing consulting work and adding it to your résumé.

The following are additional guidelines for writing a résumé:

—Customize the résumé for each employer. This means emphasizing skills that match the description of the job and the company.

—List major equipment, software, and hardware with which you have worked.

—Represent experience and technical expertise realistically.

—Never lie on a résumé.

—Do not list reasons for termination or leaving a job. Employers will discuss this in the interview.

—Do not list hobbies, sports, social activities, philosophies, or personal information about family, health, etc. The résumé should focus completely on employment skills, education, and job experience.

—If you have a college degree, do not include a high school graduation date.

—Do not state "References Available on Request" on the résumé. The employer assumes you will provide that information at the appropriate point in the interview process.

—Do not use professional jargon unless you are certain that the résumé will be read by people who understand it.

—Ask a faculty member or a person from the college placement office to proofread the résumé and offer advice.

—Print résumés on a laser printer or other high quality printer. If you don't have access to such a printer, take it to a shop that specializes in printing for a small fee.

—Print the résumé and cover letter on light colored (white, light gray, or off-white) stationary-quality paper.

—Mail the letter and résumé in a matching envelope.

## Scanner-Friendly Résumés

**Examples of Key Words**

AAS Degree. Analysis. Atomic Absorption. Balances. Chemistry. Sampling. Log. Titration. CFR. Chromatography. CPR. Environmental Technology. EPA. Eudora. Excel. First Aid. HAZCOM. HAZWOPER. Lotus. Monitoring. Wastewater. Netscape. OSHA. Spectrophotometer. Training. QA/QC. Quality Control. Word. WordPerfect. 40-hour.

Large companies are now scanning résumés into computer databases so they no longer have to handle, sort, and store the hundreds of résumés they receive. To narrow their interview selections to lists of résumés that more closely match job descriptions, employers use software programs to search résumé databases for key words. Résumés containing those key words are printed in a report for the employer to review. To increase the odds of being pulled out of the database, study the company's job description or classified advertisement for words associated with the position and use all of them in your résumé. To make the résumé "scanner friendly," consider adding a "Key Words" section to the bottom of the résumé listing all of the key words mentioned in the résumé separated by periods or commas. In addition to the key words, include industry jargon associated with the skills and experience that you possess.

The following are pointers for preparing a résumé for scanning:

—Use a plain type font such as Helvetica.

—Printed the résumé on white bond paper using a laser printer to ensure the sharpest possible print.

—Used a font size between 10 and 12 points.

—Eliminate the use of italics, boldface, underlines, and parentheses.

—Removed shading, graphics, boxes, and lines.

—Printed only your name on the first line.

Suzie Barnes
789 West Street
Willow Hills, Texas 77631
(409) 555-6543

January 11, 1999

Brobane Waste Systems
100 Hummingbird Lane
Willow Hills, Texas 77631

To Whom it May Concern:

I am writing in response to your advertisement for an Air/Water Sampling Technician that was posted with the Placement Office at Westhills College on January 10.

As a December 1998 graduate of the Environmental Technology program at Westhills, I have completed courses in regulations, sampling and analysis, industrial processes, and hazard communications. In addition, I have current 40-hour HAZWOPER, confined space entry, and first aid/CPR certificates.

Through the hands-on training in the program, as well as experience while interning at Johnson Analytical Labs, I am familiar with standard sampling protocol. I am also trained to operate the majority of field equipment associated with air, soil, and water sampling and analysis. The combination of hands-on training and current certifications should allow me to begin work as a sampling technician with a minimum amount of additional training.

Brobane Waste Systems has earned a reputation for creative and innovative solutions to waste management, and I would enjoy working for such a forward thinking organization. If you have any questions regarding my résumé, please feel free to contact me at 555-6543.

Sincerely,

Suzie Barnes

Figure 7-3: Example of a Cover Letter

### Job Placement Companies and On-Line Services

In some cases, job placement companies and on-line services can distribute your résumé to employers to which you would not normally have access. This is especially true if you are willing to relocate. Placement companies get paid only if you get hired, so they tend to serve as your advocate. In addition, job shops can be valuable sources of information about employers and the specific skills for which they are looking.

The following is a list of current on-line job services. Note that on-line job services can be bulletin boards for viewing listings, posting résumés or linking to a job placement company. Note, also, that on occasion URLs change their locations.

| | |
|---|---|
| America's Job Bank | http://www.ajb.dni.us |
| Careermosaic | http://www.careermosaic.com |
| Careerpath.com | http://www.careerpath.com |
| Environmental Careers Bulletin | http://ecbonline.com |
| Environmental Careers Organization | http://www.eco.org |
| Electronic Résumé | http://www.resumix.com |
| Get A Job!!! | http://www.getajob.com |
| Jobcenter | http://www.jobtrak.com |
| The Monster Board | http://www.monster.com |

## Cover Letters

Résumés should always be accompanied by a cover letter printed on the same bond paper as the résumé or, for e-mail, precede the résumés. Cover letters should follow the letter (for mail or FAX) or memo (for e-mail) formats discussed in Chapter 3 and include the following:

—A brief explanation of why you are contacting the company

—A summary of your skills and how they will allow you to function within the position

—A personalized statement

—Special information about how you can be reached

In cover letters mention the job title, how you learned of the position, and a summary statement about why you are qualified for the job. Avoid grandiose statements that overstate your qualifications or potential value to the organization. Next, write a personalized statement, which is a description of why you would like to work for the company.

Note that the initial impression of your cover letter will likely determine if the résumé is read. A cover letter should be free from spelling, grammatical, and typographical errors. These errors send the employer the message that you do not devote adequate attention to details.

## Checking Your Understanding

### Activity A

Write or discuss responses to the following questions:

1. How would you develop a résumé for a compliance officer position compared to a résumé for a sampling technician position?
2. What messages does a misspelled word or grammatical error in a résumé send to an employer?
3. Assume that you are currently unemployed due to a knee injury. Knowing that many employers will not hire a worker that has been laid off due to an injury, what activities could you participate in while recovering to minimize the gap in your work history?

### Activity B

Think about what type of environmental health and safety position you would like. Write a job objective for that position and develop a résumé tailored to that particular position.

### Activity C

Convert Figure 7-2 into a scanner-friendly résumé. Be sure to include a "Key Word" section.

## Activity D

From a newspaper, select a job advertisement that you would consider and prepare a cover letter to accompany a résumé. Start by explaining how you learned about the job, list relevant academic training, prior work experience, extracurricular activities, and, if appropriate, your availability and contact information. When describing your experiences and abilities use action verbs. For example, if you worked in a grocery store, write that you were responsible for *maintaining* product inventories on the shelves, *preparing* seasonal displays, and *supervising* others. Finally, explain why you are well qualified for the job and why you would like to work for them.

## Activity E

Select a company that is advertising for help in a local newspaper and prepare an *unsolicited* letter of application for an environmental health and safety (EH & S) position. Research the company by either calling for a brochure or visiting their Internet site. Obtain the name of the person to whom you should address the letter. Start the letter with a simple statement about the kind of job your are seeking. Briefly, write about your education, job experience, and the type work you are qualified for and interested in doing.

# 7-4 Preparing for Interviews

## Concepts

- Studying a company, developing a list of informed question, and practicing basic interview questions will increase the likelihood of a successful interview.
- Employers should only ask job- and skill-related questions. They cannot ask about applicants' marital status, pregnancy, religion, race, or country of origin.
- A neat, clean, and conservative appearance and a professional, friendly manner are critical in job interviews.

Congratulations! Your résumé was effective, and you have been asked to interview for the job. Now what? Employers tell stories of applicants showing up for interviews in all manner of attire from bathing suits to grubby work clothes, accompanied by everyone from small children to their mothers. They tell about applicants who argued with them, refused to talk during the interview, smelled bad, and displayed annoying habits such as smacking gum. Needless to say, these individuals are probably still looking for a job.

Interviewing does not have to be an excruciating experience for either you or the interviewer. In fact, some people actually enjoy the process because they find the experience helps them to polish their interview skills and expand their job-search network. In preparation for an interview, learn as much as you can about the company. Questions you may want to research include:

—How is the business owned? Is it family-owned, a partnership, a public agency, a non-profit, or owned by stockholders?

—Who is the president of the company, and who is the department manager where you would work?

—What are the company's products or services?

—What is the philosophy of the company?

—How old is the company?

—How many people are employed by the company, and what is the average turn over rate (i.e., how long do employee stay)?

—How much profit does the business generate each year?

—Who is the competition?

Large companies are easier to research than small because they have extensive Internet sites that provide information about the company and its products, shareholders, and profit margins. If you cannot locate an Internet site, then call the company's public relations office and ask for an information packet. You can also research large companies through publications such as *The Wall Street Journal* and *Business Week*, trade magazines, and professional journals. A librarian can be very helpful in locating information.

Small companies usually do not have public relations offices, but many of them are beginning to take advantage of the opportunities offered by the Internet. If this does not yield any information, call the company directly. Often the receptionist can provide the desired information. Other sources that may prove useful are the local Chamber of Commerce, an instructor, or a member of the industry advisory committee. Even talking to classmates may be helpful because someone may know a person who is currently working for the company. This process of sending out feelers is called networking, and it can result in some very valuable contacts.

While researching a company look for information that will aid you in anticipating some of the interviewer's questions and in formulating intelligent questions about the business or job. Just the fact that you have taken the time to read about the business sends the following messages to the interviewer:

—The job interview is important to you

—You want the job

—You are self-motivated

Research may yield information that will make you eager to work for a company - or it may not. Even if you learn something negative that convinces you not to consider a company as an employer, go through with the interview. Any interview experience will polish your skills and increase confidence. You might even learn something else about the company to change your mind.

## Interview Questions

In preparing for an interview, take a few minutes to think about your responses to the following common interview questions:

Q. Why did you choose this field?
A. Emphasize the positive. Perhaps you really liked chemistry or enjoyed the detail and structure of compliance work. Perhaps there was a particular person you admired who piqued your interest in the field. Avoid responses that indicate you are only going into the field because of the hours, money, or benefits.

Q. What do you like most about the work?
A. Use this question to emphasize your skills and show your enthusiasm for the work. If a task or group of tasks is particularly satisfying to you, say so and state why.

Q. What do you like least about the work?
A. If you really like all aspects of the job, say so. If not, answer the question as positively as possible. One way is to name what you don't like and then show how you have tried to improve your understanding of that area. For example, it would be unwise to say something like "I hate giving presentations." A better response might be "I was really nervous when I first began making presentations, but I knew that to be successful in this field, I had to be comfortable talking in front of large groups. Over the last year, I joined Toastmasters to improve my speaking skills and have begun to enjoy giving talks and making presentations."

Q. What are your strengths?
A. Describe the skills that would be most useful in the position for which you are applying. Use this question to emphasize these strengths and other skills. For example, if you are prompt, dependable, and goal-orientated, say so.

Q. What are your weaknesses?
A. Respond as positively as possible, but do not exaggerate and say "I don't have any weaknesses." The interviewer will expect an answer because, as we all know, no one is perfect. On the other hand, do not say, "I'm really lousy with details." No employer wants an employee that can't pay attention to details. A better response might be, "I have to pay extra attention to details to make sure that I have recorded information correctly and have not left anything out. To make sure that I don't make mistakes, I plan my work carefully so that I have adequate time to do the job right, and I always check my work before I submit it."

Q. Why did you leave your last job, or why do you want to leave your current job?
A. In general, graduating students have an easy response to these questions as they are usually leaving a part-time job or a job outside their field to pursue full-time employment. However, some students have events in their job histories that can make it difficult for them to answer this question. In the past, being laid off was equivalent to being fired in the minds of many interviewers. Given the amount of downsizing that has occurred in corporate America over the last few years, interviewers are now more understanding. If this is the case, put a positive spin on the situation. You might say, "I was laid off when Bowman Manufacturing was bought out by Spillman. I had always wanted to go to college. I did some research on job opportunities in the area of sampling and analysis, and found that some of the skills I already had were applicable to that field. I decided to use the retraining benefits that Bowman offered to become an environmental technician."

Q. What did you like most/least about your former boss?
A. This is a loaded question, so think carefully about how to answer it. Regardless of which way this question is asked, emphasize former employers' good points. For example, if a former boss was too detail oriented say, "Mr. Smith was extremely knowledgeable in the field of environmental compliance, and I learned a great deal from him. I learned to take extra care when I researched compliance issues for him and always double-checked my sources. Deadlines were important to him so I learned to plan ahead and be very careful about scheduling my work."

Q. May I contact your current employer for a reference?
A. If your current employer knows you are looking for a new job, be ready to provide the interviewer with his name, telephone number, and address. If your current employer does not know you are looking for another job, explain that to the interviewer and ask her to give you an opportunity to notify your supervisor. Most interviewers will respect this request. If you just say "no" without explanation, the interviewer will assume you have something to hide.

In addition to the above questions, the interviewer may ask about other items in your résumé. Put yourself in the employer's place and think about what you would ask if presented with your résumé. Anticipate the interviewer's questions and think of positive responses. Take the time to write down responses to these questions. Practice the answers so you will recall the important points even if nervous during the interview.

Finally, take a few moments to think of some questions that you can ask the employer. Write the questions down, and take them with you to the interview. You probably won't need to ask all of them because other questions will occur to you. During an interview it is appropriate to ask for details about the following:

—The duties you would be performing

—The manufacturing processes used

—The equipment you would be using

—The size of the department

—The company's professional development requirements

—What the interviewer sees in the future for the company

It is inappropriate to ask questions about salary, benefits, vacation, etc. Such questions are only relevant if a job has been offered and may leave the impression that you are only interested in the money. Always avoid controversial topics. Why the previous employee left is not a good topic to discuss during the first interview. If you must ask about a situation that is potentially embarrassing, ask as delicately as possible.

## Responding to Inappropriate Interview Questions

As noted in Chapter 6, the Equal Employment Opportunity Commission (EEOC) prohibits employers from discriminating against a potential employee based upon race or color, age, national origin, pregnancy, religion, or disability. This means that if applicants have the skills to perform the job for which they are applying, then employers cannot refuse to hire them because of one of these conditions. Consequently, an interviewer cannot ask certain questions during the course of the interview or place them on the job application. Interview questions should be directed toward the applicant's educational background, previous work experience, and technical skills. The following are questions employers can and cannot ask:

—If you have an obvious disability, or you disclose one during the course of the interview, then the employer has the right to ask you to explain or demonstrate how you will perform the job duties. However, the interviewer cannot ask you about the extent or specific nature of your disability.

—The employer has the right to ask you to submit to a physical exam if you have been offered the job and all other employees in that position or similar positions have been required to have a physical as well.

—Drug addiction is not considered a protected disability, but the guidelines as to what is an appropriate question are unclear. Some employers avoid the drug use issue by adopting a blanket drug testing policy for all employees. Under those circumstances, the employer has the right to ask the applicant if they are willing to submit to a drug test after the position has been offered. Employment is then contingent upon the applicant passing the drug-screening test.

—Interviewers should not ask about an applicant's marital status, pregnancy, religion, race, or country of origin.

Despite EEOC regulations, some employers still ask inappropriate or illegal questions. Before making a decision about how to respond, reflect on the following questions:

—What was the context in which the question was asked?

—What is the seeming intent of the question?

—Were similar questions found on the application or asked at other times during the interview?

If the question appears to be asked out of innocence or if it occurred during the informal talk before or after the interview, you may choose to answer the question and avoid any confrontation. If you are unsure about the intent, or if the same type of question occurred several times during the interview process, you may choose to call it to the interviewer's attention by stating that the question is inappropriate or illegal and decline to answer. It should be noted, however, the employer may perceive this as an indication that you would be a problem employee. In these circumstances, you will not likely get the job, but it may not be much of a loss. If the employer willfully violates the law during an interview, then there is a chance that this pattern is prevalent throughout the operation of the business.

If discrimination is suspected, you have the option of requesting that the EEOC investigate the hiring practices of the business. For more information on filing claims, access the EEOC Internet site at http://www.eeoc.gov/facts/.

## Personal Appearance

First impressions are extremely important, and the way you dress speaks volumes about you before you say a word. Personal appearance can make the difference between getting and not getting a job.

To some extent, the job will determine what is proper attire. When applying for a field position more casual dress is acceptable; such as dress pants and a nice shirt or sweater. For desk jobs or managerial positions, the required dress is more formal. Generally, clothing should be conservative, neat, pressed, and clean. Shoes should be conservative and polished. Jewelry should be kept to a minimum, such as a watch and a wedding or class ring. If in doubt about the type of clothes you should wear, talk with your college job placement office or a faculty member.

Pockets should be empty of jingling coins, etc. Do not chew gum or smoke during the interview, even if the interviewer smokes or offers you a cigarette. It is acceptable to bring a light briefcase containing a pen, résumé, transcript, portfolio, list of prepared questions, and photocopies of pages from your college catalog showing program degree plan and course descriptions.

Finally, make sure to practice good personal hygiene such as bathing, washing hair, cleaning nails, and brushing teeth. Hair should be a natural color (as opposed to blue or green), freshly cut, and neatly styled. For men, facial hair, such as beards and mustaches, is not allowed in many industrial setting because of the need to wear respirators.

## Making the Best Impression

On the day of the interview, arrive well in advance. It is unacceptable to arrive late. If you are unsure of the location of the interview, call and ask directions. Then visit the location a day or two before the interview. In downtown locations, parking is often a problem. Locate a public parking garage in advance and be sure to bring parking money.

Do not drink alcohol or smoke before the interview. Employers will not hire an applicant if they smell alcohol and are becoming reluctant to hire smokers. Due to the increasing cost of health insurance, employers are hesitant about adding anyone to their group coverage who could potentially increase expenses.

Upon arrival you may be asked to fill out an application. Using your résumé as a reference, fill out the application completely, leaving no blank spaces. When introduced to the interviewer, smile and act genuinely glad to have the opportunity to compete for the job. Be certain to remember the interviewer's name. Address the interviewer as Mr. Smith or Ms. Jones. Shake hands firmly and make eye contact. Sit only after the interviewer offers you a seat. He or she will probably begin by giving a brief overview of the company and the job. If you have done your homework, you will probably already know much of this information. Even so, look interested and pay attention. Place your hands in your lap and don't cross your arms because the interviewer may interpret this body language as being closed to the information presented. Maintain eye contact throughout the interview.

When the interviewer begins asking questions, avoid responding with one or two word answers. Answer the questions as completely and honestly as possible. Make an effort to put a positive slant on responses and look for places to emphasize your skills and knowledge. Remember that the name of the game is to make sure you have left the interviewer with a complete understanding of your skills and strengths. Be careful not to interrupt the interviewer. Do not argue under any circumstances.

The question and answer portion of the interview is an excellent time to show a portfolio. Begin with a brief explanation that it is a collection of projects and work completed in courses or related jobs. Turn the portfolio toward the interviewer and guide him to the sections you want to highlight, allowing enough time for the information to be absorbed. Show examples that illustrate a particular skill. For instance, if the interviewer is concerned that you have never worked in the field, then point to examples of field notes completed during an internship or project in school. As the interviewer flips through the material, sit quietly.

If a plant tour is conducted during the interview, be sure to observe all safety regulations, which means wearing safety glasses and a hard hat without having to be reminded. Observe walkways and caution signs. If you do not know what specific safety regulations are in force, ask. As you meet people out on the units, smile, shake hands, and talk with them. They may tell you something about the job they are working on or explain a portion of the manufacturing processes. This is an opportunity for you to show off your interpersonal skills, so be open and friendly.

At the end, thank the interviewer for his or her time and be sure to send a thank-you note (see Chapter 3). Writing a note not only shows good manners, but also brings you and your résumé to the mind of the interviewer one more time.

One last thought to consider is that the interview process is as much about the interviewer finding a good employee as you finding a good job. During interviews observe how employees interact. Do they seem to enjoy their work? Is the work area safe and well maintained? How do supervisors and other employees interact? Do the employees you talk with seem to take pride in what they are doing? Is the work environment relaxed and casual or is it more formal? Is it an environment in which you would be comfortable? If the answers to these questions are not favorable, then think long and hard about the job before accepting it, particularly if there are major safety problems.

Searching for a job can be a long and tough process. Your job search may take longer if you are unwilling or unable to relocate. In such a case, remain positive. Make getting a job your full-time employment, which means work at it every day. During the interim, consider taking additional classes or doing volunteer work in a related field. If you cannot find a job in the department you want, but like the company, consider taking another position and applying for a transfer when an environmental technician job becomes available.

# Checking Your Understanding

## Activity A

Prepare written responses for each of the questions under the Interview Questions heading in this section.

## Activity B

Select, research, and write about two local companies where you would consider working. Talk with their human resource personnel and find out how the company advertises positions and selects applicants. Develop a written plan on how you would attempt to get a job at these companies. If, for example, a company files all their openings on an Internet site or with a local employment agency, then the plan might include checking the Internet site regularly or talking with an agent at the employment agency. Be prepared to present your paper to the class.

## Activity C

It is not uncommon for individuals to relocate or accept a job that requires a fairly long commute. Determine the longest acceptable commute. Take a map, and using a compass, draw a circle with a radius equal to that distance. Develop a list of potential employers within the circle. Begin to systematically research what job openings they have and their hiring practices. If ready to start working, customize your résumé for each potential employer and begin applying. Don't forget to include city, county, and state agencies as well as hospitals in the list.

## Activity D

1. Write four general questions you can ask an interviewer. The exact wording of each question may need to be modified from company to company, but it should identify an area of particular interest such as on-the-job training opportunities.

**Application for Employment**
**Boseman's Brownfields Assessment & Redevelopment Co.**

| | | | |
|---|---|---|---|
| Name: | | | |
| Address: | | | |
| City/State/Zip: | | | |
| Phone: | | FAX: | |
| Date of Birth: | | Marital Status: | |
| Race: | Sex: ❑ M | ❑ F | |
| **Education:** | | | |
| School/Address | Dates Attended | Degree Earned | Area of Study |
| | | | |
| | | | |
| **Previous Employers:** | | | |
| Employer/Address | Dates Employed | Job Duties | Reason for Termination |
| | | | |
| | | | |
| **Certifications:** | | | |
| Certification | | Date Earned | Date Expired |
| | | | |
| | | | |
| **Reference Contacts:** | | | |
| Name | Phone Number | Employer | Address |
| | | | |
| | | | |
| **Community and Professional Affiliations:** | | | |
| Hobbies: | | | |
| Have you ever been convicted of a crime? | | ❑ No ❑ Yes Provide details in the space below. | |
| Have you ever been injured on the job? | | ❑ No ❑ Yes Provide details in the space below. | |
| Have you ever collected unemployment benefits? | | ❑ No ❑ Yes Provide details in the space below. | |
| Do you have any physical or mental disability or illness that would hinder yourabilityto perform the job you are applying for? | | | |
| ❑ No ❑ Yes Provide details in the space. | | | |
| I certify that the information listed above is accurate to the best of my knowledge. Boseman's Brownfields Assessment ant Redevelopment has my permission to contact previous employers and references listed on this application | | | |
| Signature: | | Date: | |

An Equal Opportunity/Affirmative Action Employer

Figure 7-4: Fictitious Application for Employment Form

2. Write a basic thank-you letter that can be customized and sent after each interview. For help with this activity, refer to Chapter 3.

## Activity E

Review and revise the fictitious application in Figure 7-4 to ensure that it meets EEOC standards.

## Activity F

Read Scenario 1 in Appendix 1.

One strategy that companies are using to minimize discrimination is to prepare a set of position-specific interview questions to reduce the risk of inappropriate or illegal questions. These questions are then asked of each applicant.

Locust® Co., a small manufacturing operation, is hiring a new Environmental Health and Safety Technician. The individual hired will be responsible for environmental health and safety training at the facility; developing a hazard communications plan, a storm water run-off plan, and a hazardous materials contingency plan; and overseeing the general safety and environmental policies of the company. Locust® Company has a mandatory physical exam and drug-testing program. A selection committee will do the interviewing.

In class, separate into groups of four to six students and develop 10 to 15 position-specific questions for applicants. Remember that the questions should follow EEOC standards and focus on determining if the applicant has the skills to perform the job.

# Summary

This chapter explains the integral parts of the job application process: the portfolio, cover letter, résumé, and the interview. The basic guidelines for the preparation of these three documents and interviewing techniques includes the following:

—Developing a portfolio that contains a collection of work that highlights the skills learned and indicates the level at which you are capable of performing.

—Writing a cover letter that provides a personalized introduction of you and your résumé to a potential employer.

—Developing a résumé that is clearly written, neat in appearance, well organized, and emphasizes the positive aspects of your skills and employment history.

—Developing interviewing skills by researching the company and preparing responses to commonly asked interview questions.

PRESERVING THE LEGACY

# 8

# Interpersonal Skills

## Chapter Objectives

Upon completing this chapter, the student will be able to:

1. **Analyze** customer service "moments of truth."
2. **Inventory** listening skills and set goals for self-improvement.
3. **Identify** the responsibilities of a team member.
4. **Recognize** the qualities of an effective team leader.
5. **Explain** strategies for resolving issues before they become a source of conflict.
6. **Explain** strategies for resolving issues after they have become a source of conflict.

## Chapter Sections

# 8-1 Importance of Interpersonal Skills

## Concepts

- Interpersonal skills are the skills used for getting along with others.
- Interpersonal skills are valuable to employers.
- The interpersonal skills of employees can attract or drive away customers during their first contact with a company.
- Customers can be internal or external.

As mentioned in Chapter 1, employers rank communication and teamwork skills equally with technical expertise – especially for entry-level positions. Good communication and teamwork are dependent on employees' interpersonal skills, which add up to the ability to get along with others.

Businesses today recognize two types of customers:

—**External customers** who are the traditional customers outside the company such as clients, purchasers, and vendors.

—**Internal customers** who are the people within a company, such as colleagues in the same department or people in other departments to whom services are provided. For example, all employees are customers of payroll departments because payroll departments provide the service of issuing checks. In turn, all payroll departments are customers of department supervisors who provide the service of submitting accurate time sheets.

Author Ron Zemke writes about "moments of truth," in *The Service Edge*. Such moments include when a customer assesses the value of doing business with a company upon first contact with an employee. For the customer, this initial contact defines the company. If a customer is treated courteously and effectively, then he will be inclined to use the company's services or product. If the customer is, for example, ignored or given poor service then his actions will be the opposite. Therefore, whoever has first contact with a customer is at least initially responsible for winning or losing business, regardless of whether the employee is a first-line worker or a manager. This is just one reason why employers place a high value on interpersonal skills.

Getting along with customers and fellow employees does more than create a pleasant atmosphere; it affects the quality and quantity of your work and that of everyone around you.

> "...only 4% of unhappy customers will complain about a problem while the other 96% will switch to a competitor."
>
> Ron Zemke,
> *The Service Edge*

An entire body of literature about interpersonal communications exists in libraries and bookstores. The purpose of this chapter is to motivate you to learn more about the subject. You may want to start by looking at this book's Bibliography for book titles.

## Checking Your Understanding

### Activity A

Develop a log of your "moments of truth" for the next week, ranking each first contact with a business as either a positive or a negative experience. For example, was the clerk at the supermarket pleasant and helpful? Did the registrar or counselor solve a scheduling problem cooperatively?

### Activity B

1. Select three of the situations from the log in Activity A and develop a table such as the one that follows for each. Under the center category, write a list of expectations that you had for the situation. Next, write what actually happened under Experience Enhancers and/or Experience Detractors. An example of a Moment of Truth table[1] is as follows:

[1] Zemke, Ronald with Schaaf, Dick, *The Service Edge*, Plume Publishing, 1990, page 36.

| Moment of Truth<br>A student presents a scheduling problem<br>to the college registrar | | |
|---|---|---|
| **Experience Enhancers** | **Standard Expectations** | **Experience Detractors** |
| — Example: The registrar appeared to understand my situation and knew what to do.<br>— ... | — I will only have to see the registrar once to solve my problem.<br>— I will be able to talk to the registrar without a long wait.<br>— The registrar will take an interest in my problem.<br>— The registrar will be pleasant.<br>— The registrar will explain exactly what will happen next. | — The registrar acted as if it was a bother when I presented my situation.<br>— ... |

2. Share your tables with the class and participate in finding common "experience enhancers" and "experience detractors." For example, some students may have encountered people who were unwilling to try to solve their problems.
3. Form several small teams and develop a list of "Customer Service Do's and Don'ts." Discuss the lists developed by the different teams.

## Activity C

1. In class, separate into small groups to come up with several internal and external customer service situations that could be encountered by an environmental technician. An example:

   While taking carbon monoxide readings in the field, an external customer approaches a technician and asks if he could stop by his house on the way home to check a hot water heater vent for possible carbon monoxide leaks.

2. For each situation, provide suggestions about how it could be handled in a positive manner.
3. In teams, report the examples and suggestions to the class.

# 8-2 Listening Skills

## Concepts

- Listening involves information processing and human relations.
- Listening skills are critical for keeping customers satisfied.
- Interrupting, finishing sentences, and constantly asking for the information again are signs of a poor listener.

Listening skills tend to fall into two broad categories: information processing and human relations. Information processing involves the receiving of information regarding some plan, procedure, fact, or instruction. Human relations are the emotional elements, or feelings, that can filter or sometimes block communication.

> "Listening to customers must become everyone's business. With most competitors moving ever faster, the race will go to those who listen (and respond) most intently."
>
> Tom Peters, *Thriving on Chaos*

The possible consequences of poor listening on the job are a dangerous situation, a costly error, an offended co-worker, or a customer lost. When working with customers, environmental technicians (particularly those in business for themselves) must listen carefully and understand what is being said for the following reasons[2]:

—**To understand customers' "moments of truth."** What are the customers' "critical contact points" and how effective is each point of contact?

—**To keep track of market trends.** What are the current needs and expectations of your customers? Are you set up to serve their needs and expectations?

—**To hear customers' unexpected ideas.** A new idea could increase business.

—**To involve the customer in the business.** Involving customers builds business relationships that last.

> "The average listener retains only half of what is said in a ten-minute oral presentation. Within 48 hours, that drops by half again. As a result, only about 25 percent of what is heard is understood, evaluated, and retained. Even worse, ideas may be distorted by as much as 80 percent as they travel through the chain of command in an organization."
>
> *Effective Human Relations in Organizations, p. 57, quoting Susan Mundale, "Why More CEO's Are Mandating Listening and Writing Training," Training/HRD, October 1980, pp. 37-41.*

To determine if you are an effective listener, ask yourself the following questions:

1. Do you often interrupt the speaker?
2. Do you tend to finish sentences for other people?
3. Do you frequently ask to have things repeated?

If you answered yes to any of the above questions, then you need to reevaluate your listening skills. You may need to practice thinking about what is being said, rather than about what you are going to say next. Resist distractions and pretending to listen as your mind wanders. In short, make a conscious effort to learn to pay better attention when others are speaking.

## Listening Skills Evaluation

The following questions are designed to help you evaluate your listening skills. Use your answers to pin point problem areas in your listening habits.

1. Do you actively work at listening?
2. Do you deliberately tell yourself to listen?

[2] Zemke, Ronald with Schaaf, Dick, *The Service Edge*, Plume Publishing, 1990, page 29.

3. Do you take notes when someone gives you instructions and then repeat the instructions for clarification?
4. Do you ask questions when you are not sure you understand?
5. Do you keep your mind open to new ideas and opinions that differ from yours?
6. Are you able to set aside emotions when you are supposed to be listening?
7. Do you give verbal (such as "I understand" or "I see") and nonverbal (nodding your head) feedback to the speaker to encourage him to continue?
8. Do you focus on the content of what is said?
9. Do you tune out the speaker's appearance or mannerisms if they threaten to get in the way?
10. Are you patient?
11. Do you avoid finishing other peoples' sentences?
12. Do you reduce environmental distractions (e.g., holding calls or moving to a quieter location)?
13. Do you keep good eye contact while listening?
14. Do you avoid mannerisms such as tapping a pen while the speaker is talking?
15. Do you put down your work and turn your full attention to the speaker?

If your answer to any of these questions is "no," then try to change your listening behavior in that area. Do not expect improvement to occur over night. Long-held habits are always difficult to break, but the reward is that you will become a better listener and have stronger interpersonal skills.

## Checking Your Understanding

### Activity A

For the next week, use the listening skills evaluation in this section to assess your listening skills. Write summaries of your strengths and weaknesses, and set two goals for self-improvement.

### Activity B

**Scenario**

One day you are in a client's office doing routine paperwork. The alarm system begins its special buzz pattern that indicates a chemical spill. Immediately you see workers filing out in a quick, orderly way as they evacuate the plant. You make your way to the designated incident command post and see two hazardous materials technicians suiting up in respirators and protective clothing so they can approach the spill area. They will use two-way radios to report what they see to the incident commander. The incident commander will assess their observations and direct them on how to proceed. In a sense, the technicians will become the eyes and ears of the commander. Therefore, it is critical that the technicians communicate effectively to the incident commander and that the commander carefully listens to and understands the technicians.

As the consultant who trained all of these personnel on how to respond to a spill, you are now curious as to how they will function in this actual emergency. After the emergency reaches its conclusion, you call the technicians and incident commander together for a debriefing to evaluate the experience.

In class, simulate the scenario described above.

**The actors** should include three students playing the roles of the two hazardous materials technicians and the incident commander. The rest of the class is the audience.

**The props** required are a two-way radio, flip chart and pad, marker, tape measure, and spill material (e.g., corn meal, dry beans, or water).

**The staging area** of the incident scene must be out of view of the incident commander. Before the simulation takes place, several students should create a "spill" around an area that adds interesting features to the accident such as next to machinery, flowing under a door, or close to a drain.

**Perform the first activity.** Using two-way radios, simulate the technicians describing the spill area to the incident commander using terms that clearly communicate the situation. As the commander interprets their communication, he sketches the scene on a large flipchart sheet. The commander asks for clarification about the scene until he and the technicians feel that they have successfully communicated and the sketch is finished.

**Perform the second activity.** Role-play the debriefing session where all class members, including the hazardous materials technicians and the incident commander, evaluate the effectiveness of the communication during the incident. Begin by taking the sketch to the scene and determining how closely the drawing matches the actual area. The class members can then discuss ways to improve communication, such as strategies for orienting the drawing (e.g., "at 12 o'clock, I see such and such; at 2 o'clock, I see such and such...") Additional debriefing questions could include the following:

—Were there any barriers to communicating?

—Did the technicians use any particularly effective strategies for communicating the scene?

—What did the incident commander say to obtain clarification?

—What did class members learn from their observations of the role players?

**Other activities** include using the ideas from the first debriefing on another spill situation with the goal of increasing accuracy while decreasing the time spent in communicating.

## Activity C

While variations are widely used by workshop trainers, the main purposes of the Building Block exercise are team development and communication skills enhancement. In this exercise several teams compete with each other to reproduce a model structure out of sight from each team's "building site." The winning team builds the best copy of the model in the shortest time.

**The materials and the setting** include the following:

—A large number of building blocks grouped into identical sets. One set is the model. The other sets are the building materials that each team must use to copy it

—Separate rooms or large cardboard boxes in front of each of the models. Ideally, the model should be in one room, and each team should have a separate room to build their copies. If the exercise must take place in a single meeting room, then find a way to provide visual barriers, such as the cardboard boxes.

**Team Assignments** include the following. Note that each team can have one or more people fulfill these roles:

Builder — The Builder stays at the building site to construct the copy according to verbal instructions passed along by the Runner.

Runner — The Runner, without ever seeing the model or the copy, delivers verbal instructions from the Looker to the Builder. At the request of the Builder, the Runner delivers descriptions of the needed parts to the Supplier and returns blocks to the Builder.

Looker — The Looker looks at the model and gives the Runner instructions to pass along to the Builder.

Supplier — The Supplier listens to the Runner's description of the part, selects the appropriate part, and gives it to the Runner who delivers it to the Builder.

Overseer — The Overseer may look at the model as well as the copy to help clarify situations for the team. However, the Overseer is limited to saying "That's right" or "That's wrong."

**The first activity** is the contest itself. Before this begins, the instructor tells the class how long the teams have to try to duplicate the original block configuration. The instructor then monitors the contest, notifies the class when time is up, and declares the winning team.

**The second activity** is an evaluation and discussion session where the class, either in groups or in an open discussion, answers the following questions:

—Which group was the most successful in replicating the model?

—What factors contributed to their success?

—What were the difficulties encountered and the steps taken to overcome those difficulties?

—What improvements should be made?

—What is the most important skill that people should have to work effectively together when performing different tasks?

—What strategies can team members employ to help distinct divisions of labor (such as those in the exercise) work effectively together?

—How can the duties of the looker, builder, runner, overseer, and supplier be related to jobs found in the business world?

An example of an answer to the last question could be as follows. During an overhaul of a generator, an engineer (analogous to the Looker) writes specifications and gives them to a contractor (the Builder). The contractor then interprets the work that needs to be done and sends a list of required parts to a supplier. The supplier contracts with a trucking company (the Runner) to deliver the parts. The customer (the Overseer) tells the builder if the generator is being built as he intended.

# 8-3 Teamwork Skills

## Concepts

- The two main types of teams are job oriented and project oriented.
- Teams usually use the same process for completing a project, which includes understanding and defining the mission, brainstorming, researching, implementing, and evaluating.
- A mission statement encapsulates the team's goals into one sentence.
- Teams produce better quality work than one person because they bring together diverse talents and perspectives.
- For a team to be a success, team members must understand the team's goal(s), maintain a positive attitude, and deliver the work required.
- Team leaders must have strong interpersonal, administrative, and communication skills.
- Team leaders must be responsible for cultivating a sense of belonging among team members, recognizing the contributions of individual members, and providing team support.
- The steps to changing a disruptive member of the team's behavior are diagnosing the problem, describing the problem, reaching an agreement, coaching, and reinforcing.

In the workplace, teams are either job- or project-oriented. Members of a job-oriented team work in a department together, such as a team of field sampling technicians who work together on a daily basis. A project-oriented team is a group of employees brought together for the expressed mission of completing a single project. An example of a project-oriented group is a safety committee, where representatives from different departments work together to implement a safety program.

Regardless of the type of team, participation within a team gives members a feeling of ownership toward their assignments, their jobs, and the business as a whole. Members of teams work together to consider a wide range of solutions or strategies for solving a problem and then systematically narrow those strategies to select the most viable one. In addition, teams are usually involved in the implementation of the strategies or solutions and the evaluation of the results.

> "Teamwork is the starting point for treating people right. Most people think that teamwork is only important when competing against other teams, but competition is only part of the picture. In most things we do in life, people have to work with rather than against each other to get things done."
>
> Dave Thomas
> Founder of Wendy's restaurant chain

Few rules exist for team size, composition, or assignments. A team may be formed to work on a short- or long-term project. The projects may involve development, creation, management, and maintenance. For example, a team of environmental technicians may be asked to design a system that makes MSDSs available to employees, implement the system, and ensure that the system is maintained and improved. Team members could be recruited because of their backgrounds or volunteer out of interest in the assignment. In general, teams are most effective when management's interference is minimal, the organizational climate is free of fear, and the suggestions and conclusions made by the team are followed. Note that while management's participation should not be overly intrusive, they normally monitor the project's implementation and evaluation processes.

### How Do Teams Work?

Teams use similar methods to complete a job or a project. The main steps that teams use to approach projects are shown in Figure 8-1 and described as follows:

—**Understand and define the mission** It is important that all members understand the task the group is expected to perform. The team's mis-

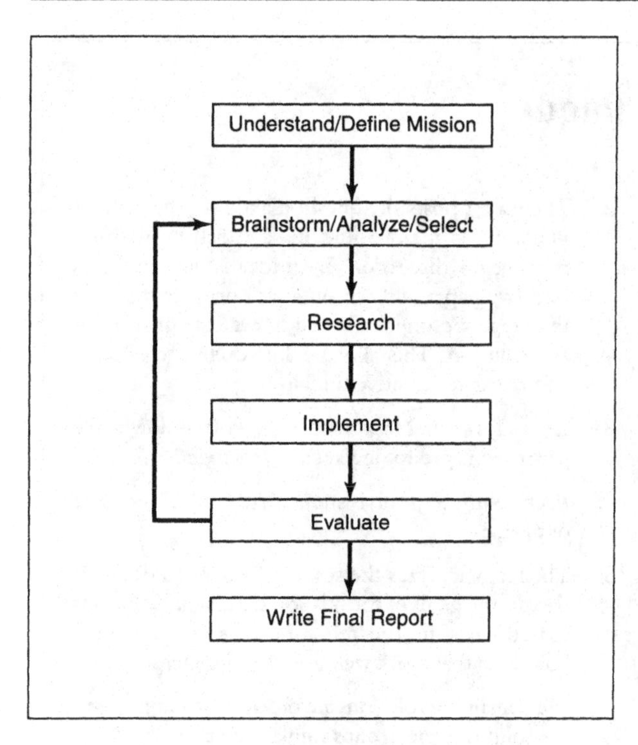


Figure 8-1: Steps Teams Use to Approach Projects

sion should include constraints within which the group must work, such as budgetary limitations. For example, the implementation of a mission to identify waste streams in a production process and to employ pollution prevention alternatives will differ between a budget of $100,000 and $500,000. One of the best ways to ensure that each person on the team understands the mission is to have the group write a mission statement. For example, a first attempt at a mission statement might be "To prevent pollution using a budget of $100,000." As the group looks at the statement, they might want to further define what type of pollution they want to prevent and how the $100,000 budget is to be allocated. After consulting with management, the team might change the mission "To reduce the amount of water pollution using an annual budget of $100,000."

—**Brainstorm** The objective of brainstorming is to generate a wide range of possible solutions or recommendations. Creativity is encouraged, and no ideas are rejected at this stage. Next, the ideas generated during brainstorming sessions are analyzed. The list of ideas is then narrowed, and the team selects the best option.

—**Research** Brainstorming solution(s) should be researched to confirm their feasibility. For example, one of the solutions "To reduce the amount of water pollution using an annual budget of $100,000" could be the purchase of a reverse osmosis treatment system. The team would then have to research the prices to purchase and maintain such a system.

—**Report to Management** The mission statement and the team's solution are documented for management's approval.

—**Implement the Plan** Before implementation takes place, tasks are assigned and deadlines are set. The team leader monitors the progress and reports to management. Adjustments are made as necessary.

—**Evaluate the Results** The team analyzes how well the implementation accomplished the mission statement's goals and documents their finding for management. If the solution is not effective, the team could try the second solution on their brainstorming list or return to brainstorming and use what they have learned to generate better ideas. To find the best solution, the group could cycle through the brainstorming-selecting-implementing-evaluating process several times.

—**Write the Final Report** After the solution has been satisfactorily implemented, a final report is written that documents the processes the team has worked through. The value of this report is that it provides a road map, summarizes research, and records the reasons for successes and failures so new team members or later teams will not have to go through the same efforts.

In parallel with the work process described above are the stages a team's group dynamics go through before members work effectively together. Typically, they are the following:

—An introductory phase where participants assess other members in the group

—A brainstorming phase in which members become more comfortable with one another, and the group begins to search for a strategy to tackle the task

## Box 8-1 ■ A Brainstorming Technique

**Materials:**

— Adhesive note pads (4 x 6 in.) and free wall space or a large table

— Marker pens

**Ground Rules:**

— All ideas are considered.

— No idea is criticized.

— No person is criticized.

**Method:**

1. The team changes the mission statement into a question. For example, "To reduce the amount of water pollution using an annual budget of $100,000" becomes "How can our company reduce the amount of water pollution with an annual budget of $100,000?"
2. The leader posts the question and asks the team to generate as many solutions as possible without pausing for discussion. Each idea is written on an adhesive paper and put on a wall or large table. At this stage, no one is allowed to evaluate or rule out any solution. This allows a free exchange of ideas and enhances creative thinking.
3. After all of the ideas have been written down, the participants group similar solutions together.
4. Each solution group is summarized in one sentence or phrase.
5. The team analyzes the solution groups to decide if they fit the goals of the mission statement. Solutions that do not fit the mission goals are removed. Solutions that work well together are merged.
6. The remaining solutions are ordered, beginning with the solutions the groups thinks are better.
7. The team keeps removing, merging, and prioritizing until the best solution or set of solutions emerges. Throughout, the team constantly refers back to the mission statement.

—A resolution phase where the group narrows down a wide range of options identified during the brainstorming stage into a specific course of action

—The implementation phase where the team begins executing, evaluating, and correcting a plan of action

Occasionally, groups struggle during first three phases. If a group becomes stranded at one of these stages, they will be unable to complete the assigned task.

## Advantages and Disadvantages of Teams

One of the greatest advantages of teamwork is summed up in the saying, "Two heads are better than one." Each team member brings a unique perspective to bear on a project, and each of those perspectives provides information that allows the group to explore innovative solutions. Other advantages include the following:

—Team members can balance one another's strengths and weaknesses. For example, a creative person's weakness of not tending to details could be compensated for by an analytical individual who enjoys working on the fine points of a project. The creative individual who quickly becomes bored with a project can spur the perfectionist to wrap up the details and proceed to the next step in the process.

—Teams allow the work and the responsibility of large projects to be spread over several individuals, thereby reducing stress. Even when difficult decisions must be made, team members can draw consolation from the fact that several individuals share in the responsibility for that decision.

—Team members tend to encourage and support one another. While unforeseen setbacks and delays can discourage an individual worker, team members tend to encourage disheartened members, thus keeping the project moving forward.

—When working alone, it is easy to become distracted or to procrastinate over a difficult segment of a project. Regular meetings to check the team's progress keep members on track because they know that their progress is monitored by their peers.

—Team members tend to take greater responsibility for a project.

—Team members tend to feel a greater degree of job satisfaction. This is largely due to the fact that teams members are given the opportunity to add their input and have a sense of control over the environment in which they work.

Disadvantages include the following:

—Working in teams can be slower because a greater number of perspectives must be considered.

—Poorly formed teams, or teams headed by a weak leader, can lose direction or become mired in details.

—Conflicts within a team can stifle progress.

—Some individuals dislike working in teams because they feel that the opportunities for individual recognition are less than when a single individual is responsible for the entire project.

Despite these potential difficulties, the creativity and quality of work that teams produce is well worth the effort.

## Team Members

Members of a team must meet certain obligations. First, team members are responsible for having a complete understanding of the team's mission. If a team member cannot complete the sentence, "This team was assembled to __(goal of the team)__," then he should study the relevant materials or talk with the team leader. It is difficult to reach a goal or solve a problem if you don't understand it. Other team member responsibilities include the following:

1. Doing what is required to see the project through to completion
2. Performing necessary research
3. Keeping an open mind
4. Listening carefully to other members
5. Contributing your strengths
6. Accepting assignments and completing them in a timely and thorough manner
7. Notifying the team leader in advance if a deadline cannot be met so other arrangements can be made or other team members can assist

One of the most important contributions to any team is a good attitude. For team members to make worthwhile contributions to a project, they must demonstrate through their actions that the project is of value. This attitude is communicated by the commitment that members display. Committed individuals are on time and prepared for meetings, and they deliver on their assignments. They do not gripe or whine.

Effective team members also display an attitude of acceptance toward other members' ideas and contributions – particularly when the team is working through a brainstorming session. The true strength of a team lies in its members' ability to explore a variety of perspectives and solutions regarding a problem. Ridiculing and criticizing ideas put forth by other team members stifles the creative process. Sometimes this is done by members stating that an idea is stupid or won't work. Other times, it is done in less obvious ways, such as ignoring someone's opinions or talking while a team member is trying to communicate ideas.

Critiques and analyses should be focused on the merits of the idea, not the person who offered it. Comments like "Jane's idea won't work because she doesn't know anything about marketing" or "Remember what happened the last time we tried one of Harry's schemes" demonstrate a negative attitude toward team members and should be avoided. If a team member thinks that an idea is not the solution for the problem, then he should state the reasons and back them with data and diplomacy. "We tried that before and it didn't work" may be a valid reason to discount an idea, but can also lower team morale. A responsible team member would support that opinion with documentation showing that the idea had been implemented and an analysis of why the idea failed.

Finally, team members must be flexible, cooperative, and willing to compromise. They may sometimes have to perform tasks that do not play to their strengths, yet still must be done. Effective team members accept their assignments and complete them to the best of their ability. Even if a solution is not a first choice to all of the team, each member must be willing to compromise. Perhaps experience will bear out the concerns of the dissenting team members, but those team members also have a responsibility to refrain from the "I told you so!" syndrome. It is at those times that team members should be willing to regroup and reenter the brainstorming-selection-implementation-evaluation process to come up with a more effective method of meeting their goal.

## Team Leaders

If appointed or chosen to become a team leader, you will become the person most accountable for the completion of a project and play a variety of roles. The main responsibilities of a team leader are as follows:

—Ensuring the effectiveness of the team

—Monitoring the progress of the team in reaching their goal

—Acting as a communication liaison between the team and management

To accomplish the above, a team leader needs a variety of skills. Leaders must have highly developed interpersonal skills to:

—**Cultivate a sense of belonging among team members.** Helping individuals feel that they are an important part of the group may include simple things such as scheduling meetings when all members can attend and being sure that all members are notified. A more creative method of elevating members' feelings of importance may be promoting team camaraderie by designing team T-shirts or arranging time outside of work for members to get to know one another better.

—**Recognize the contributions of individual members.** Identifying a job well done prevents members from feeling that being on a team lacks opportunities for individual recognition. Team leaders can provide recognition by drawing the group's attention to a particularly creative idea or well-done research, writing notes to upper management, or crediting members with the role that they played in the project.

—**Provide team support.** Perhaps most importantly, team leaders must offer support to members as they work in the team. This may include recognizing that the employee needs some additional time off after meeting a tight deadline, providing additional training, or negotiating with management for additional resources if the project proves larger than originally expected.

Along with interpersonal skills, team leaders must have strong administrative skills, as they must perform the following duties to further assure the effectiveness of their teams:

—Mediating and helping team members resolve conflicts

—Ensuring that meetings are held regularly and setting clear agendas for them

—Delegating the workload

—Clearly communicating the expected quality of work

—Setting realistic deadlines and taking appropriate action when they are missed or ignored

Finally, leaders must be good communicators as they are liaisons between management and the team, conveying the concerns and desires of one group to the other using written and oral communication. Team leaders can keep management and their teams working together by listening carefully to each side and accurately communicating information.

## Uncooperative Team Members

Uncooperative team members exhibit a variety of bad habits, such as the following:

—Missing meetings and deadlines, forcing others to pick up their slack

—Criticizing to the point of abusing other team members

—Refusing to cooperate, only accepting the choicest assignments, and refusing to contribute on other more mundane tasks

—Grandstanding, stealing the limelight from the rest of the group

—Refusing to accept change, fighting the integration of new or creative solutions

—Back stabbing and sabotaging the efforts of the team

Regardless of the behaviors exhibited, uncooperative behavior can damage the effectiveness of the entire team. While individual team members might be able to moderate the behavior of an uncooperative member, it is the group leader's responsibility to correct it. The following are the basic steps for dealing with an uncooperative team member:

—**Diagnose the problem** A leader should never make assumptions about someone's behavior. He should always ask the person directly what is causing the destructive behavior. The team member's reasons may be legitimate. For example, people can become uncooperative when their workload is excessive. If the complaints are reasonable, then adjustments need to be made. If not, then the next step should be taken.

—**Describe the problem** If the individual's complaints or problems are unfounded, then the leader describes the effect the member's behavior is having on the group. Sometimes, particularly in the case of a negative or overly critical team member, he may not realize the effect his behavior is having on the group.

—**Reach an agreement** After the person is made aware of his behavior, he must agree on adhering to a standard of conduct that is acceptable in the group.

—**Coach and reinforce** After reaching an agreement, the group leader may have to coach the individual and reinforce acceptable behavior. Coaching might include providing administrative support to help the employee meet his or her team obligations. For example, the group leader might arrange for additional training if a lack of instruction is contributing to the problem. Reinforcement includes expressing approval of the improved behavior and helping the person integrate back into the group by making other team members aware of examples of improved behavior.

Unfortunately, not all individuals are willing to correct their behavior. In such cases, the only effective course of action is removing the individual and replacing him with someone willing to participate in the group's activities. When a leader is unwilling to take this strong course of action, a message is sent to the team that there is lack of support for them. The result is morale suffers and members may become less committed to the team's goal. Team members may also perceive that their behavior is inconsequential – that it's all right to miss deadlines, disrupt meetings, and argue. Usually, the disruption caused by keeping a problem team member is far greater than having him reassigning to another task.

# Checking Your Understanding

## Activity A

1. Make a list of what you think your strengths and weaknesses as a team member would be. List your strengths in one column and weaknesses in another. Consider a variety of skill areas by asking the following:

   —Am I a good diplomat who can help different factions "make peace" with one another?

   —Do I think creatively?

   —Do I write or proofread well?

   —Do I tend to discount other people's ideas or be overly critical?

2. Once your list is completed, write a paragraph or two discussing how you could use your strengths in a team. Discuss what skills your fellow team members would need in order to compensate for your weaknesses.

## Activity B

As an Environmental Health and Safety (EH & S) technician, you have been assigned to a team whose mission is to identify pollution prevention strategies. A new EH & S technician, Jay Harmon, is also assigned to the team. Since he is new, he doesn't have any knowledge of the company's history. As a result, his ideas are regularly shot down by the rest of the team with phrases like "Management will never go for it" or "We tried that and it didn't work." Write a paragraph or two answering the following questions:

1. As a team member, how could you help integrate Jay into the team?
2. What benefit would Jay's lack of knowledge regarding company history be to the group?

## Activity C

### Scenario

As a community service project, Wall Production has agreed to teach members of a grassroots organization how to gather air and water samples. Since the grassroots group is without major funding, Wall Production will make a $1,000 cash donation and provide a team of Wall employees to handle the following:

—Planning the training

—Soliciting additional funds and equipment donations from other industry partners

—Advertising the training sessions to the local press

### Activity

In class, divide into groups, with each group portraying the Wall employee team. Using a budget of $1,000, each team must develop a strategy for creating training materials, soliciting donations, and advertising training sessions. Note that the presentations must include information about how the workload would be divided. After each group presents their implementation plan, the class critiques it by answering the following questions:

—Has the team addressed all pertinent points?

—Has a realistic timeline been developed?

—Is the team's implementation plan detailed and thorough?

—Has the $1,000 been used wisely?

—What issues has the team failed to consider?

—Does each team member appear to be contributing to the overall goal of the group?

—Was the presentation clear, concise, and well organized?

# 8-4 Conflict Resolution

## Concepts

- Conflicts can seriously disrupt productivity and lower moral.
- The best way to deal with conflicts is to prevent them from happening.
- If a dispute has occurred, the best strategy is to address the problem openly with all of the parties involved.
- Flaws in work processes can cause conflict among employees.

Since employees depend on one another for success, disputes can seriously disrupt productivity. Conflict erodes the professionalism of the company. It also obstructs businesses because more energy is channeled toward managing internal conflict than beating the competition. Minor disputes between two or three employees can disrupt or impair an entire department. Even when productivity is not seriously affected, conflict destroys the general harmony of the workplace and undermines employee moral.

The best ways to deal with conflicts are preventing them from happening and resolving them with tact and diplomacy when they do.

## Conflict Prevention

Since conflict often arises from misunderstandings or from actions that erode the trust among workers, effective strategies for reducing conflict are those that promote understanding and build trust, such as the following:

—Monitor your own behavior; that is, keep your word, be dependable, meet deadlines, and be considerate of other individuals' workloads and deadlines.

—Use listening skills to decrease confusion; for example, ask questions when you do not understand a policy or process.

—Avoid guessing a person's intent. Do not read more into a situation than is actually there. Again, ask questions to clarify the situation.

—Be generous with your praise and diplomatic with your criticism.

—Thank fellow workers who go out of their way to assist you.

—Do not gossip. If you would not say something to a person face-to-face, then you should not say it behind his back.

—Be tolerant of co-workers. We all make mistakes, so focus on correcting the errors rather than blaming others.

—Avoid circumstances that make you irritable. For example, if looming deadlines make you irritable, then complete the work ahead of schedule.

## Conflict Resolution

Despite prevention, conflicts occasionally arise. Once a dispute has occurred, the best strategy is to address it, preferably by opening a dialogue with those involved, and do the following:

—Use "I-messages" when bringing up issues that are bothering you.

—Avoid an accusing tone of voice. The other employees may not realize that what they are doing is creating problems for you, or you may find that there are circumstances that explain the dispute, such as in the automotive parts case study (see Box 8-2)

—Analyze your actions and evaluate whether they contributed to the problem.

—If you were wrong, then admit it and apologize.

---

An "I-message" consists of these parts:

— *I feel*...(state your feeling)

— *when* (describe behavior).

For example, "*I feel frustrated when you are frequently absent from work.*"

---

## Box 8-2 ■ Case Study

Disputes between workers hinder the resolution of problems, as described in the following case of an automotive part manufacturer.

When assembly line workers began packing shipping crates only half-full of certain automotive parts it created problems for several related departments. Shipping encountered problems because the cost of delivering parts increased dramatically. Customer Support found that since parts could not be securely braced within the half-full crates, customers were complaining about damaged products. Employees in Shipping and in Customer Support were saying among themselves that they thought the assembly line workers were incompetent.

The gossip finally got around to the assembly line workers, who responded that they had been advised to fill the crates half full by management. A change to a heavier part resulted in full crates being overweight, so management instructed the workers not to pack crates beyond a certain weight limit. The assembly line workers were simply following orders.

Unfortunately, Shipping and Customer Support had spent several weeks criticizing the work of another department rather than solving the problems of increased shipping costs and goods damaged in shipping by requesting a redesign of the shipping crates.

Sometimes compromise is necessary to let another individual save face. Consider the following example:

> Don, a very bright student interning at an environmental engineering firm, was assigned to Mark, who did not have a college degree. Over a period of years, Mark had worked his way up from a laborer's job to a management position. The internship had appeared a perfect match for Don's abilities and interests, but after three weeks he requested permission to change internship hosts. Don complained that he just could not seem to please Mark. Mark complained that Don was a burden and, after all of Don's classes, he should not need to ask so many questions. On the other hand, when Don tackled assigned tasks without asking questions, Mark belittled him for mistakes in front of the rest of the department. When the supervising faculty member contacted Mark about the problem, Mark complained about being forced to "baby-sit" some intern and about what a "know-it-all" student he was. It became clear that the real problem was that Mark found Don's enthusiasm threatening. Mark was well aware that most people in his position had at least an associate degree. He interpreted the fact that he had been assigned an intern as an indication that management was considering replacing him with an employee who had a degree. In Mark's opinion, the internship was just an excuse for management to compare his skills with those of a college graduate.
>
> While Don was not guilty of anything more than youthful enthusiasm, it was clear that if the internship was going to work, the student needed to look for ways to help Mark save face. Together, Don and the supervising faculty member devised some ways to work with Mark in a less threatening manner. For example, Don stopped using phrases like, "In ENV235 we took samples using..." because Mark might take that to imply that since he had not had ENV235, he must be unqualified to be the manager. Don began to phrase questions in such a way that he deferred to Mark's years of experience in the position. It took several weeks for Mark's feelings of distrust to subside, but eventually the two were able to work comfortably together, and Don learned a great deal, both in terms of technical skills and interpersonal skills.

Once a course of action to resolve a conflict has been developed and implemented, then follow up to be sure the compromise is working. If it is not working, review what actions have been taken so far, and consider other alternatives.

Often, parties in a dispute can find ways to compromise. However, there are circumstances where you should never compromise. If asked to do something that is against company policy, illegal, or dangerous, always stand your ground. Explain why you feel the compromise is unacceptable. Be prepared to cite specific policies or laws, and, if an alternative compromise cannot be devised, immediately involve an appropriate supervisor.

A final option for resolving disputes is reviewing what happened and improving the underlying system or process in which the conflict occurred so that the problem does not reoccur in a slightly different form. This is done by finding the root cause of a dispute, identifying where it takes place in a process, and working with the people involved to correct it. For example, in the case study the root cause of the conflicts was poor communication. Both of the situations could have been prevented if the managers had reported their directives to other department managers, or the workers had made the changes in procedures known to their colleagues in other departments.

## Checking Your Understanding

### Activity A

Write a paragraph describing a possible solution to the following scenario:

The Locust® plant manager, Mr. Turf, launches into a tirade about the new emergency response training that you spent weeks developing for the Environmental Health and Safety (EH&S) department. You wonder if Mr. Turf is really angry with you or the fact that the training schedule, decided by the company's owner, Mr. Crabgrass, overrides a much needed vacation. The training comes on the heels of a major equipment overhaul that has kept Mr. Turf and his crew working 12-16 hour shifts for the last three weeks. At the moment, he is tired, unhappy about the prospect of canceling his vacation, and not looking forward to listening to the complaints of his overworked crew. Normally, Mr. Turf is a reasonable person. You suspect that under different circumstances, he would probably be receptive to the new training.

### Activity B

**Scenario**

G. B. Adams recently assembled a team of employees to guide the implementation of an ISO 14000 Environmental Management System at his small plastics extruding company. The group met initially and elected Carol Sams as leader. Because of the amount of work associated with the implementation process, Mr. Adams gave Carol a $500 per month stipend for the overtime she will be spending heading the group.

The $500 stipend created problems within the team since none of the other members were given extra pay. Consequently, some team members took over the delegation responsibilities and assigned Carol more tasks than she could ever hope to complete. Every time Carol requested assistance, the team sat silently.

With the work grinding to a halt and meetings degenerating into gripe sessions about how little progress the chair was making on her "assigned" tasks, Carol decided to confront the group. She scheduled a meeting, but two days beforehand the president of the company called to express his displeasure with the team's progress. At the end of the call, Mr. Adams was adamant that the company could not pay each team member a $500 stipend and that the work must be completed since ISO 14000 certification is required to bid on lucrative contracts.

The team members are as follows:

—**Carol Sams** has worked for Adams only about six months, but her creative solutions to a number of design problems quickly brought her to the owner's attention. Chairing the ISO 14000 team is a position that Carol would like to retain because, once ISO certification is obtained, a new ISO management position will be created. The head of the implementation committee is the logical person to fill the new position.

—**Jack Wilks** has been working for the company for twenty years and is resentful that someone so new to the company has been appointed chair of this important group. He also thinks Carol should not have been paid any extra money. Jack is well educated and has years of design experience, but he does have a hearing problem. He has been unwilling to get a hearing aid and his deafness accounts for his often bizarre or inaccurate responses during discussions.

—**Paul Mailis** is a process operation technician from the production unit and has worked for Adams for two years. He is pleased that the company is seeking ISO 14000 certification, because he feels it will make the company more competitive in a global market. He would help more with the tasks at hand but isn't exactly sure what he can contribute. He is not resentful that Carol is getting additional money because he has done little work as of yet.

—**Bob Morgan** is the manager of shipping and receiving. He has a reputation for being a "good ole' boy" and a real terror to work for. His behavior is often erratic. Minor issues have been known to send him into a screaming rage, and his subordinates tiptoe around him at all times. He sees ISO 14000 as an unnecessary task and attends the meetings only because Mr. Adam has personally requested that he join the team.

—**Craig Gomez** is the senior manager of the accounting department and is very pleased that the Adams Co. is pursing ISO 14000 certification. He has some experience with ISO 14000 from working on a similar committee for his previous employer. He is resentful that Carol is getting additional money.

—**Kara Pimberton** is Vice President of Research and Development. She's not sure that Adams Co. has the resources to achieve ISO 14000 certification but supports the idea nevertheless. She empathizes with Carol and tries as best she can, given her busy schedule, to provide support.

Pretend that you are Carol Sams and write a paper on how you would use the upcoming meeting to get your team on the right track. Think about what you have learned about teams and conflict resolution. To get started, consider the following questions:

—What should your goal be during the meeting?

—What difficult points do you need to resolve before the group can proceed?

—Each team member has specific reasons for wanting or not wanting the ISO 14000 certification implemented. How can you capitalize on those that do?

—What talents and resources do you have within the team?

—Should you continue accepting the stipend or voluntarily give it up?

—Are some team members so adversarial to the group that they should be removed? Who are they and how should they be removed?

Share your paper in class. Then, either as a class or in groups, create an agenda for the next meeting.

## Summary

Interpersonal skills pertain to how we interact with others. Such skills can affect a company's quality of service or products. Customer's "moments of truth," (those good or bad first impressions that determine whether customers will continue to do business with a company) are often the result of an employee's interpersonal skills.

Two categories of listening skills are information processing and human relations. Effective listening strategies include paying attention to the speaker's message, focusing, taking notes, clarifying by asking questions, keeping your mind open, giving feedback, reducing distractions, and maintaining eye contact.

The team approach to achieving job- and project-oriented goals is a process that includes:

—Understanding and defining the mission

—Brainstorming/analyzing/selecting

—Researching

—Implementing

—Evaluating, which may reveal the need for a team to cycle back to brainstorming and to move through the process again

The advantage of a team completing a project is that the result is usually better than what could be created by one person. This is because each team member brings a unique perspective and set of talents. The disadvantages are that working in teams can be a slower process, conflicts can occur, and opportunities for individual recognition are fewer.

The success of a team depends on the team members and the leader. Team member must understand the team's goal(s), maintain positive attitudes, and deliver the required work. Team leaders must possess strong interpersonal, administrative, and communication skills.

Prevention is the best way to avoid conflict in the work place. This is done by promoting understanding and trust.

If a dispute has occurred, strategies include using "I-messages," voiding an accusing tone of voice, analyze your actions, admitting if you were wrong, and apologizing. A final step in resolving disputes is to improve the process under which the problem occurred.

# Appendix 1

## Scenario 1

You have recently completed an environmental technology associate degree at a local two-year college. Locust®, a medium-sized company that makes commercial lawn mowers, has hired you to be their Environmental Health and Safety Technician. The company's primary manufacturing operations involve cutting, welding, painting, and the assembling of lawn mowers.

Part of your first day on the job is spent with the company owner, Mr. Crabgrass. He explains that this is a new position because Mr. Green, the Human Resources Director, could no longer wear two hats. Mr. Crabgrass also explains that although the company has a written Hazard Communication (HAZCOM) program, Mr. Green had prepared it by using one that the Human Resources Director at another company had provided. His secretary retyped it and put in the Locust® Company name. Mr. Crabgrass recently heard from a golfing friend that the company should have a storm water run-off plan as well as a hazardous materials contingency plan. He admits that he is uncertain about what these are or if they are necessary. He is even more uncertain about whether there are other written plans they might need. He says that this is why you were hired, and it is one of the things he wants you to report on as soon as possible.

Mr. Crabgrass then introduces you to Arnold Turf, the plant manager. As you walk around with Arnold to become familiar with the manufacturing operation and people, you note that safety glasses and ear protection devices are hanging on doorknobs or sitting on bench tops, instead of being worn. You observe a cutting operation where the safety guard has been removed from the machine, apparently to perform routine maintenance on its working parts, and not replaced. The noise level prevents you from mentioning this to Arnold. You also note that several of the operators have rigged temporary auxiliary lighting by draping power cords over their machines. One of the painters empties a spray gun and then rinses it with solvent and pours it into a nearby floor drain. A worker applying a powerful smelling adhesive to the slip-proof mat on the operator's deck is also eating a candy bar. You note a number of poor housekeeping practices that are adding to the amount of hazardous waste being collected. For example, the collection drum for oily rags also contains old newspapers and candy bar wrappers.

As you complete the first day and are walking to your car, you can't help but wonder how many other unacceptable environmental health and safety practices you will discover. The thought is hardly finished when you see the custodian dribbling a dark liquid along the joints in the pavement. You ask, "What are you doing to the joints?" "Oh," says the custodian, "we have found that a mixture of waste oil and paint thinner works wonders for preventing weeds and grass from growing in these cracks."

## Scenario 2

You have been hired as a technician by Environmental Perfection Investigation and Consulting (EPIC). Your job is to assist EPIC's environmental engineers. They provide the following types of services:

— Air and water monitoring

— Site assessments

— Regulatory compliance

— Process reviews and good housekeeping practices

— Record keeping and reporting

— Waste audits

— Assessment and selection of personal protective equipment

You and one of the engineers have been assigned to work with a new client, Unified Diversified Limited (UDL), which is a small manufacturer of automotive instrument panels. It employs 184 production workers, 13 office staff members, 18 maintenance workers, and 7 managers/supervisors. N.U. Dashboard is the owner and founder of the company. Bill Kleen is the safety director. Marty Willfix is the maintenance manager. UDL's primary product is an integrated automobile dashboard that includes all electrical components, LCD display panels, air bags, heating-ventilation-air conditioning (HVAC), and audio components. The one-piece design features a single wiring harness connector and HVAC port, which greatly simplifies its installation on the assembly line. This dashboard has become very popular with auto manufacturers because it bends. This flexibility makes it easy to install or remove, if repairs of any of the dashboard components are ever needed.

The basic dashboard is produced using a high-pressure injection molding process, followed by a patented cold milling technique. To complete the design of the soft, molded plastic dashboard, it is cooled to –40°C and milled to design specifications using a computer assisted milling (CAM) operation. When the milling is complete, the unit is warmed to room temperature, degreased with a hot alkaline cleaner, rinsed in deionized (DI) water, and dried. Both the alkaline cleaner and rinse water is released to the sewer. Copper-silver foil is then bonded to the panel to provide for electrical connectivity of the various components. After the application of a masking agent, nitric acid is used to etch the circuitry into the foil. The excess acid is rinsed from the panels and allowed to enter the sewer. After drying, UDL ships the panels to Mexico where each of the components are mounted and the final wiring completed. The Mexican plant then ships the completed dashboards back to UDL for quality control testing and distribution to the automotive assembly plant located in Kentucky.

UDL is a safety-oriented company. They have material safety data sheets (MSDSs) for all chemicals used in the plant. Hazard communication (HAZCOM) and spill response training for all employees have been kept up-to-date. The plant is well organized, with hazardous materials storage in a designated outbuilding and bulk chemical storage and ammonia refrigeration units located in a service hall that is away from the production floor. The outbuilding and the plant service hall are also equipped with safety in mind. Each area contains the following:

— Appropriate and maintained safety equipment

— OSHA, EPA, and DOT shipping placards and labels

— MSDSs for all chemicals

— Posted process diagrams and piping layouts

As a part of their goal to achieve ISO 14000 certification, UDL has contracted with EPIC to make their good plant practices even better. ISO 14000 is a set of industrial management standards created by the International Organization for Standards. These standards are designed to help organizations address environmental issues in a systematic way and thereby improve their environmental performance. ISO 14000 certification is often required for bids on industrial contracts. You and the environmental engineer will be part of UDL's study team, which also includes management and first-line workers. Together the team will analyze every environmental safety and health aspect to determine the areas of excellence and those that could show improvement. In keeping with the intent of ISO 14000 certification, the results of this study will serve as the focal point for UDL's goal of continuous environmental improvement.

# Appendix 2

## OSHA Regulations (Standards - 29 CFR) Hazard Communication - 1910.1200

**Standard Number: 1910.1200**
Standard Title: Hazard Communication.
SubPart Number: Z
SubPart Title: Toxic and Hazardous Substances

(a) **"Purpose."**

(1) The purpose of this section is to ensure that the hazards of all chemicals produced or imported are evaluated, and that information concerning their hazards is transmitted to employers and employees. This transmittal of information is to be accomplished by means of comprehensive hazard communication programs, which are to include container labeling and other forms of warning, material safety data sheets and employee training.

..1910.1200(a)(2)

(2) This occupational safety and health standard is intended to address comprehensively the issue of evaluating the potential hazards of chemicals, and communicating information concerning hazards and appropriate protective measures to employees, and to preempt any legal requirements of a state, or political subdivision of a state, pertaining to this subject. Evaluating the potential hazards of chemicals, and communicating information concerning hazards and appropriate protective measures to employees, may include, for example, but is not limited to, provisions for: developing and maintaining a written hazard communication program for the workplace, including lists of hazardous chemicals present; labeling of containers of chemicals in the workplace, as well as of containers of chemicals being shipped to other workplaces; preparation and distribution of material safety data sheets to em-

ployees and downstream employers; and development and implementation of employee training programs regarding hazards of chemicals and protective measures. Under section 18 of the Act, no state or political subdivision of a state may adopt or enforce, through any court or agency, any requirement relating to the issue addressed by this Federal standard, except pursuant to a Federally-approved state plan.

(b) **"Scope and application."**

(1) This section requires chemical manufacturers or importers to assess the hazards of chemicals which they produce or import, and all employers to provide information to their employees about the hazardous chemicals to which they are exposed, by means of a hazard communication program, labels and other forms of warning, material safety data sheets, and information and training. In addition, this section requires distributors to transmit the required information to employers. (Employers who do not produce or import chemicals need only focus on those parts of this rule that deal with establishing a workplace program and communicating information to their workers. Appendix E of this section is a general guide for such employers to help them determine their compliance obligations under the rule.)

(2) This section applies to any chemical which is known to be present in the workplace in such a manner that employees may be exposed under normal conditions of use or in a foreseeable emergency.

(3) This section applies to laboratories only as follows:

(i) Employers shall ensure that labels on incoming containers of hazardous chemicals are not removed or defaced;

..1910.1200(b)(3)(ii)

(ii) Employers shall maintain any material safety data sheets that are received with incoming shipments of hazardous chemicals, and ensure that they are readily accessible during each workshift to laboratory employees when they are in their work areas;

(iii) Employers shall ensure that laboratory employees are provided information and training in accordance with paragraph (h) of this section, except for the location and availability of the written hazard communication program under paragraph (h)(2)(iii) of this section; and,

(iv) Laboratory employers that ship hazardous chemicals are considered to be either a chemical manufacturer or a distributor under this rule, and thus must ensure that any containers of hazardous chemicals leaving the laboratory are labeled in accordance with paragraph (f)(1) of this section, and that a material safety data sheet is provided to distributors and other employers in accordance with paragraphs (g)(6) and (g)(7) of this section.

(4) In work operations where employees only handle chemicals in sealed containers which are not opened under normal conditions of use (such as are found in marine cargo handling, warehousing, or retail sales), this section applies to these operations only as follows:(b)(4)(i)Employers shall ensure that labels on incoming containers of hazardous chemicals are not removed or defaced;

..1910.1200(b)(4)(ii)

(ii) Employers shall maintain copies of any material safety data sheets that are received with incoming shipments of the sealed containers of hazardous chemicals, shall obtain a material safety data sheet as soon as possible for sealed containers of hazardous chemicals received without a material safety data sheet if an employee requests the material safety data sheet, and shall ensure that the material safety data sheets are readily accessible during each work shift to employees when they are in their work area(s); and,

(iii) Employers shall ensure that employees are provided with information and training in accordance with paragraph (h) of this section (except for the location and availability of the written hazard communication program under paragraph (h)(2)(iii) of this section), to the extent necessary to protect them in the event of a spill or leak of a hazardous chemical from a sealed container.

(5) This section does not require labeling of the following chemicals:

(i) Any pesticide as such term is defined in the Federal Insecticide, Fungicide, and Rodenticide Act (7 U.S.C. 136 et seq.), when subject to the labeling requirements of that Act and labeling regulations issued under that Act by the Environmental Protection Agency;

(ii) Any chemical substance or mixture as such terms are defined in the Toxic Substances Control Act (15 U.S.C. 2601 et seq.), when subject to the labeling requirements of that Act and labeling regulations issued under that Act by the Environmental Protection Agency;

..1910.1200(b)(5)(iii)

(iii) Any food, food additive, color additive, drug, cosmetic, or medical or veterinary device or product, including materials intended for use as ingredients in such products (e.g. flavors and fragrances), as such terms are defined in the Federal Food, Drug, and Cosmetic Act (21 U.S.C. 301 et seq.) or the Virus-Serum-Toxin Act of 1913 (21 U.S.C. 151 et seq.), and regulations issued under those Acts, when they are subject to the labeling requirements under those Acts by either the Food and Drug Administration or the Department of Agriculture;

(iv) Any distilled spirits (beverage alcohols), wine, or malt beverage intended for nonindustrial use, as such terms are defined in the Federal Alcohol Administration Act (27 U.S.C. 201 et seq.) and regulations issued under that Act, when subject to the labeling requirements of that Act and labeling regulations issued under that Act by the Bureau of Alcohol, Tobacco, and Firearms;

(v) Any consumer product or hazardous substance as those terms are defined in the Consumer Product Safety Act (15 U.S.C. 2051 et seq.) and Federal Hazardous Substances Act (15 U.S.C. 1261 et seq.) respectively, when subject to a consumer product safety standard or labeling requirement of those Acts, or regulations issued under those Acts by the Consumer Product Safety Commission; and,

(vi) Agricultural or vegetable seed treated with pesticides and labeled in accordance with the Federal Seed Act (7 U.S.C. 1551 et seq.) and the labeling regulations issued under that Act by the Department of Agriculture.

..1910.1200(b)(6)

(6) This section does not apply to:

(i) Any hazardous waste as such term is defined by the Solid Waste Disposal Act, as amended by the Resource Conservation and Recovery Act of 1976, as amended (42 U.S.C. 6901 et seq.), when subject to regulations issued under that Act by the Environmental Protection Agency;

(ii) Any hazardous substance as such term is defined by the Comprehensive Environmental Response, Compensation and Liability ACT (CERCLA) (42 U.S.C. 9601 et seq.) when the hazardous substance is the focus of remedial or removal action being conducted under CERCLA in accordance with the Environmental Protection Agency regulations.

(iii) Tobacco or tobacco products;

(iv) Wood or wood products, including lumber which will not be processed, where the chemical manufacturer or importer can establish that the only hazard they pose to employees is the potential for flammability or combustibility (wood or wood products which have been treated with a hazardous chemical covered by this standard, and wood which may be subsequently sawed or cut, generating dust, are not exempted);

(v) Articles (as that term is defined in paragraph (c) of this section);

(vi) Food or alcoholic beverages which are sold, used, or prepared in a retail establishment (such as a grocery store, restaurant, or drinking place), and foods intended for personal consumption by employees while in the workplace;

..1910.1200(b)(6)(vii)

(vii) Any drug, as that term is defined in the Federal Food, Drug, and Cosmetic Act (21 U.S.C. 301 et seq.), when it is in solid, final form for direct administration to the patient (e.g., tablets or pills); drugs which are packaged by the chemical manufacturer for sale to consumers in a retail establishment (e.g., over-the-counter drugs); and drugs intended for personal consumption by employees while in the workplace (e.g., first aid supplies);

(viii) Cosmetics which are packaged for sale to consumers in a retail establishment, and cosmetics intended for personal consumption by employees while in the workplace;

(ix) Any consumer product or hazardous substance, as those terms are defined in the Consumer Product Safety Act (15 U.S.C. 2051 et seq.) and Federal Hazardous Substances Act (15 U.S.C. 1261 et seq.) respectively, where the employer can show that it is used in the workplace for the purpose intended by the chemical manufacturer or importer of the product, and the use results in a duration and frequency of exposure which is not greater than the range of exposures that could reasonably be experienced by consumers when used for the purpose intended;

(x) Nuisance particulates where the chemical manufacturer or importer can establish that they do not pose any physical or health hazard covered under this section;

(xi) Ionizing and nonionizing radiation; and,

(xii) Biological hazards.

(c) **"Definitions."**

"Article" means a manufactured item other than a fluid or particle:

(i) which is formed to a specific shape or design during manufacture;

(ii) which has end use function(s) dependent in whole or in part upon its shape or design during end use; and

(iii) which under normal conditions of use does not release more than very small quantities, e.g., minute or trace amounts of a hazardous chemical (as determined under paragraph (d) of this section), and does not pose a physical hazard or health risk to employees. "Assistant Secretary" means the Assistant Secretary of Labor for Occupational Safety and Health, U.S. Department of Labor, or designee.

"Chemical" means any element, chemical compound or mixture of elements and/or compounds.

"Chemical manufacturer" means an employer with a workplace where chemical(s) are produced for use or distribution.

"Chemical name" means the scientific designation of a chemical in accordance with the nomenclature system developed by the International Union of Pure and Applied Chemistry (IUPAC) or the Chemical Abstracts Service (CAS) rules of nomenclature, or a name which will clearly identify the chemical for the purpose of conducting a hazard evaluation.

"Combustible liquid" means any liquid having a flashpoint at or above 100 deg. F (37.8 deg. C), but below 200 deg. F (93.3 deg. C), except any mixture having components with flashpoints of 200 deg. F (93.3 deg. C), or higher, the total volume of which make up 99 percent or more of the total volume of the mixture.

"Commercial account" means an arrangement whereby a retail distributor sells hazardous chemicals to an employer, generally in large quantities over time and/or at costs that are below the regular retail price.

"Common name" means any designation or identification such as code name, code number, trade name, brand name or generic name used to identify a chemical other than by its chemical name.

"Compressed gas" means:

(i) A gas or mixture of gases having, in a container, an absolute pressure exceeding 40 psi at 70 deg. F (21.1 deg. C); or

(ii) A gas or mixture of gases having, in a container, an absolute pressure exceeding 104 psi at 130 deg. F (54.4 deg. C) regardless of the pressure at 70 deg. F (21.1 deg. C); or

(iii) A liquid having a vapor pressure exceeding 40 psi at 100 deg. F (37.8 deg. C) as determined by ASTM D-323-72.

"Container" means any bag, barrel, bottle, box, can, cylinder, drum, reaction vessel, storage tank, or the like that contains a hazardous chemical. For purposes of this section, pipes or piping systems, and engines, fuel tanks, or other operating systems in a vehicle, are not considered to be containers.

"Designated representative" means any individual or organization to whom an employee gives written authorization to exercise such employee's rights under this section. A recognized or certified collective bargaining agent shall be treated automatically as a designated representative without regard to written employee authorization.

"Director" means the Director, National Institute for Occupational Safety and Health, U.S. Department of Health and Human Services, or designee.

"Distributor" means a business, other than a chemical manufacturer or importer, which supplies hazardous chemicals to other distributors or to employers.

"Employee" means a worker who may be exposed to hazardous chemicals under normal operating conditions or in foreseeable emergencies. Workers such as office workers or bank tellers who encounter hazardous chemicals only in non-routine, isolated instances are not covered.

"Employer" means a person engaged in a business where chemicals are either used, distributed, or are produced for use or distribution, including a contractor or subcontractor.

"Explosive" means a chemical that causes a sudden, almost instantaneous release of pressure, gas, and heat when subjected to sudden shock, pressure, or high temperature.

"Exposure or exposed" means that an employee is subjected in the course of employment to a chemical that is a physical or health hazard, and includes potential (e.g. accidental or possible) exposure. "Subjected" in terms of health hazards includes any route of entry (e.g. inhalation, ingestion, skin contact or absorption.)

"Flammable" means a chemical that falls into one of the following categories:

(i) "Aerosol, flammable" means an aerosol that, when tested by the method described in 16 CFR 1500.45, yields a flame projection exceeding 18 inches at full valve opening, or a flashback (a flame extending back to the valve) at any degree of valve opening;

(ii) "Gas, flammable" means: (A) A gas that, at ambient temperature and pressure, forms a flammable mixture with air at a concentration of thirteen (13) percent by volume or less; or(B) A gas that, at ambient temperature and pressure, forms a range of flammable mixtures with air wider than twelve (12) percent by volume, regardless of the lower limit;

(iii) "Liquid, flammable" means any liquid having a flashpoint below 100 deg. F (37.8 deg. C), except any mixture having components with flashpoints of 100 deg. F (37.8 deg. C) or higher, the total of which make up 99 percent or more of the total volume of the mixture.

(iv) "Solid, flammable" means a solid, other than a blasting agent or explosive as defined in 1910.109(a), that is liable to cause fire through friction, absorption of moisture, spontaneous chemical change, or retained heat from manufacturing or processing, or which can be ignited readily and when ignited burns so vigorously and persistently as to create a serious hazard. A chemical shall be considered to be a flammable solid if, when tested by the method described in 16 CFR 1500.44, it ignites and burns with a self-sustained flame at a rate greater than one-tenth of an inch per second along its major axis.

"Flashpoint" means the minimum temperature at which a liquid gives off a vapor in sufficient concentration to ignite when tested as follows:

(i) Tagliabue Closed Tester (See American National Standard Method of Test for Flash Point by Tag Closed Tester, Z11.24-1979 (ASTM D 56-79)) for liquids with a viscosity of less than 45 Saybolt Universal Seconds (SUS) at 100 deg. F (37.8 deg. C), that do not contain suspended solids and do not have a tendency to form a surface film under test; or

(ii) Pensky-Martens Closed Tester (see American National Standard Method of Test for Flash Point by Pensky-Martens Closed Tester, Z11.7-1979 (ASTM D 93-79)) for liquids with a viscosity equal to or greater than 45 SUS at 100 deg. F (37.8 deg. C), or that contain suspended solids, or that have a tendency to form a surface film under test; or

(iii) Setaflash Closed Tester (see American National Standard Method of Test for Flash Point by Setaflash Closed Tester (ASTM D 3278-78)).Organic peroxides, which undergo autoaccelerating thermal decomposition, are excluded from any of the flashpoint determination methods specified above.

"Foreseeable emergency" means any potential occurrence such as, but not limited to, equipment failure, rupture of containers, or failure of control equipment which could result in an uncontrolled release of a hazardous chemical into the workplace.

"Hazardous chemical" means any chemical which is a physical hazard or a health hazard.

"Hazard warning" means any words, pictures, symbols, or combination thereof appearing on a label or other appropriate form of warning which convey the specific physical and health hazard(s), including target organ effects, of the chemical(s) in the container(s). (See the definitions for "physical hazard" and "health hazard" to determine the hazards which must be covered.)

"Health hazard" means a chemical for which there is statistically significant evidence based on at least one study conducted in accordance with established scientific principles that acute or chronic health effects may occur in exposed employees. The term "health hazard" includes chemicals which are carcinogens, toxic or highly toxic agents, reproductive toxins, irritants, corrosives, sensitizers, hepatotoxins, nephrotoxins, neurotoxins, agents which act on the hematopoietic system, and agents which damage the lungs, skin, eyes, or mucous membranes. Appendix A provides further definitions and explanations of the scope of health hazards covered by this section, and Appendix B describes the criteria to be used to determine whether or not a chemical is to be considered hazardous for purposes of this standard.

"Identity" means any chemical or common name which is indicated on the material safety data sheet (MSDS) for the chemical. The identity used shall permit cross-references to be made among the required list of hazardous chemicals, the label and the MSDS.

"Immediate use" means that the hazardous chemical will be under the control of and used only by the person who transfers it from a labeled container and only within the work shift in which it is transferred.

"Importer" means the first business with employees within the Customs Territory of the United States which receives hazardous chemicals produced in other countries for the purpose of supplying them to distributors or employers within the United States.

"Label" means any written, printed, or graphic material displayed on or affixed to containers of hazardous chemicals.

"Material safety data sheet (MSDS)" means written or printed material concerning a hazardous chemical which is prepared in accordance with paragraph (g) of this section.

"Mixture" means any combination of two or more chemicals if the combination is not, in whole or in part, the result of a chemical reaction.

"Organic peroxide" means an organic compound that contains the bivalent -O-O-structure and which may be considered to be a structural derivative of hydrogen peroxide where one or both of the hydrogen atoms has been replaced by an organic radical.

"Oxidizer" means a chemical other than a blasting agent or explosive as defined in 1910.109(a), that initiates or promotes combustion in other materials, thereby causing fire either of itself or through the release of oxygen or other gases.

"Physical hazard" means a chemical for which there is scientifically valid evidence that it is a combustible liquid, a compressed gas, explosive, flammable, an organic peroxide, an oxidizer, pyrophoric, unstable (reactive) or water-reactive.

"Produce" means to manufacture, process, formulate, blend, extract, generate, emit, or repackage.

"Pyrophoric" means a chemical that will ignite spontaneously in air at a temperature of 130 deg. F (54.4 deg. C) or below.

"Responsible party" means someone who can provide additional information on the hazardous chemical and appropriate emergency procedures, if necessary.

"Specific chemical identity" means the chemical name, Chemical Abstracts Service (CAS) Registry Number, or any other information that reveals the precise chemical designation of the substance.

"Trade secret" means any confidential formula, pattern, process, device, information or compilation of information that is used in an employer's business, and that gives the employer an opportunity to obtain an advantage over competitors who do not know or use it. Appendix D sets out the criteria to be used in evaluating trade secrets.

"Unstable (reactive)" means a chemical which in the pure state, or as produced or transported, will vigorously polymerize, decompose, condense, or will become self-reactive under conditions of shocks, pressure or temperature.

"Use" means to package, handle, react, emit, extract, generate as a byproduct, or transfer.

"Water-reactive" means a chemical that reacts with water to release a gas that is either flammable or presents a health hazard.

"Work area" means a room or defined space in a workplace where hazardous chemicals are produced or used, and where employees are present.

"Workplace" means an establishment, job site, or project, at one geographical location containing one or more work areas.

..1910.1200(d)

(d) **"Hazard determination."**

(1) Chemical manufacturers and importers shall evaluate chemicals produced in their workplaces or imported by them to determine if they are hazardous. Employers are not required to evaluate chemicals unless they choose not to rely on the evaluation performed by the chemical manufacturer or importer for the chemical to satisfy this requirement.

(2) Chemical manufacturers, importers or employers evaluating chemicals shall identify and consider the available scientific evidence concerning such hazards. For health hazards, evidence which is statistically significant and which is based on at least one positive study conducted in accordance with established scientific principles is considered to be sufficient to establish a hazardous effect if the results of the study meet the definitions of health hazards in this section. Appendix A shall be consulted for the scope of health hazards covered, and Appendix B shall be consulted for the criteria to be followed with respect to the completeness of the evaluation, and the data to be reported.

(3) The chemical manufacturer, importer or employer evaluating chemicals shall treat the following sources as establishing that the chemicals listed in them are hazardous:

(i) 29 CFR part 1910, subpart Z, Toxic and Hazardous Substances, Occupational Safety and Health Administration (OSHA); or,

..1910.1200(d)(3)(ii)

(ii) "Threshold Limit Values for Chemical Substances and Physical Agents in the Work Environment," American Conference of Governmental Industrial Hygienists (ACGIH) (latest edition). The chemical manufacturer, importer, or employer is still responsible for evaluating the hazards associated with the chemicals in these source lists in accordance with the requirements of this standard.

(4) Chemical manufacturers, importers and employers evaluating chemicals shall treat the following sources as establishing that a chemical is a carcinogen or potential carcinogen for hazard communication purposes:

(i) National Toxicology Program (NTP), "Annual Report on Carcinogens" (latest edition);

(ii) International Agency for Research on Cancer (IARC) "Monographs" (latest editions); or

(iii) 29 CFR part 1910, subpart Z, Toxic and Hazardous Substances, Occupational Safety and Health Administration. Note: The "Registry of Toxic Effects of Chemical Substances" published by the National Institute for Occupational Safety and Health indicates whether a chemical has been found by NTP or IARC to be a potential carcinogen.

(5) The chemical manufacturer, importer or employer shall determine the hazards of mixtures of chemicals as follows:

(i) If a mixture has been tested as a whole to determine its hazards, the results of such testing shall be used to determine whether the mixture is hazardous;1910.1200(d)(5)(ii)

(ii) If a mixture has not been tested as a whole to determine whether the mixture is a health hazard, the mixture shall be assumed to present the same health hazards as do the components which comprise one percent (by weight or volume) or greater of the mixture, except that the mixture shall be assumed to present a carcinogenic hazard if it contains a component in concentrations of 0.1 percent or greater which is considered to be a carcinogen under paragraph (d)(4) of this section;

(iii) If a mixture has not been tested as a whole to determine whether the mixture is a physical hazard, the chemical manufacturer, importer, or employer may use whatever scientifically valid data is available to evaluate the physical hazard potential of the mixture; and,

(iv) If the chemical manufacturer, importer, or employer has evidence to indicate that a component present in the mixture in concentrations of less than one percent (or in the case of carcinogens, less than 0.1 percent) could be released in concentrations which would exceed an established OSHA permissible exposure limit or ACGIH Threshold Limit Value, or could present a health risk to employees in those concentrations, the mixture shall be assumed to present the same hazard.

(6) Chemical manufacturers, importers, or employers evaluating chemicals shall describe in writing the procedures they use to determine the hazards of the chemical they evaluate. The written procedures are to be made available, upon request, to employees, their designated representatives, the Assistant Secretary and the Director. The written description may be incorporated into the written hazard communication program required under paragraph (e) of this section.

..1910.1200(e)

(e) **"Written hazard communication program."**

(1) Employers shall develop, implement, and maintain at each workplace, a written hazard communication program which at least describes how the criteria specified in paragraphs (f), (g), and (h) of this section for labels and other forms of warning, material safety data sheets, and employee information and training will be met, and which also includes the following:

(i) A list of the hazardous chemicals known to be present using an identity that is referenced on the appropriate material safety data sheet (the list may be compiled for the workplace as a whole or for individual work areas); and,

(ii) The methods the employer will use to inform employees of the hazards of non-routine tasks (for example, the cleaning of reactor vessels), and the hazards associated with chemicals contained in unlabeled pipes in their work areas.

(2) "Multi-employer workplaces." Employers who produce, use, or store hazardous chemicals at a workplace in such a way that the employees of other employer(s) may be exposed (for example, employees of a construction contractor working on-site) shall additionally ensure that the hazard communication programs developed and implemented under this paragraph (e) include the following:

(i) The methods the employer will use to provide the other employer(s) on-site access to material safety data sheets for each hazardous chemical the other employer(s)' employees may be exposed to while working;

..1910.1200(e)(2)(ii)

(ii) The methods the employer will use to inform the other employer(s) of any precautionary measures that need to be taken to protect employees during the workplace's normal operating conditions and in foreseeable emergencies; and,

(iii) The methods the employer will use to inform the other employer(s) of the labeling system used in the workplace.

(3) The employer may rely on an existing hazard communication program to comply with these requirements, provided that it meets the criteria established in this paragraph (e).

(4) The employer shall make the written hazard communication program available, upon request, to employees, their designated representatives, the Assistant Secretary and the Director, in accordance with the requirements of 29 CFR 1910.1020 (e).

(5) Where employees must travel between workplaces during a workshift, i.e., their work is carried out at more than one geographical location, the written hazard communication program may be kept at the primary workplace facility.

(f) **"Labels and other forms of warning."**

(1) The chemical manufacturer, importer, or distributor shall ensure that each container of hazardous chemicals leaving the workplace is labeled, tagged or marked with the following information:

..1910.1200(f)(1)(i)

(i) Identity of the hazardous chemical(s);

(ii) Appropriate hazard warnings; and

(iii) Name and address of the chemical manufacturer, importer, or other responsible party.

(2)

(i) For solid metal (such as a steel beam or a metal casting), solid wood, or plastic items that are not exempted as articles due to their downstream use, or shipments of whole grain, the required label may be transmitted to the customer at the time of the initial shipment, and need not be included with subsequent shipments to the same employer unless the information on the label changes;

(ii) The label may be transmitted with the initial shipment itself, or with the material safety data sheet that is to be provided prior to or at the time of the first shipment; and,

(iii) This exception to requiring labels on every container of hazardous chemicals is only for the solid material itself, and does not apply to hazardous chemicals used in conjunction with, or known to be present with, the material and to which employees handling the items in transit may be exposed (for example, cutting fluids or pesticides in grains).

..1910.1200(f)(3)

(3) Chemical manufacturers, importers, or distributors shall ensure that each container of hazardous chemicals leaving the workplace is labeled, tagged, or marked in accordance with this section in a manner which does not conflict with the requirements of the Hazardous Materials Transportation Act (49 U.S.C. 1801 et seq.) and regulations issued under that Act by the Department of Transportation.

(4) If the hazardous chemical is regulated by OSHA in a substance-specific health standard, the chemical manufacturer, importer, distributor or employer shall ensure that the labels or other forms of warning used are in accordance with the requirements of that standard.

(5) Except as provided in paragraphs (f)(6) and (f)(7) of this section, the employer shall ensure that each container of hazardous chemicals in the workplace is labeled, tagged or marked with the following information:

(i) Identity of the hazardous chemical(s) contained therein; and,

(ii) Appropriate hazard warnings, or alternatively, words, pictures, symbols, or combination thereof, which provide at least general information regarding the hazards of the chemicals, and which, in conjunction with the other information immediately available to employees under the hazard communication program, will provide employees with the specific information regarding the physical and health hazards of the hazardous chemical.

..1910.1200(f)(6)

(6) The employer may use signs, placards, process sheets, batch tickets, operating procedures, or other such written materials in lieu of affixing labels to individual stationary process containers, as long as the alternative method identifies the containers to which it is applicable and conveys the information required by paragraph (f)(5) of this section to be on a label. The written materials shall be readily accessible to the employees in their work area throughout each work shift.

(7) The employer is not required to label portable containers into which hazardous chemicals are transferred from labeled containers, and which are intended only for the immediate use of the employee who performs the transfer. For purposes of this section, drugs which are dispensed by a pharmacy to a health care provider for direct administration to a patient are exempted from labeling.

(8) The employer shall not remove or deface existing labels on incoming containers of hazardous chemicals, unless the container is immediately marked with the required information.

(9) The employer shall ensure that labels or other forms of warning are legible, in English, and prominently displayed on the container, or readily available in the work area throughout each work shift. Em-

ployers having employees who speak other languages may add the information in their language to the material presented, as long as the information is presented in English as well.

(10) The chemical manufacturer, importer, distributor or employer need not affix new labels to comply with this section if existing labels already convey the required information...1910.1200(f)(11)

(11) Chemical manufacturers, importers, distributors, or employers who become newly aware of any significant information regarding the hazards of a chemical shall revise the labels for the chemical within three months of becoming aware of the new information. Labels on containers of hazardous chemicals shipped after that time shall contain the new information. If the chemical is not currently produced or imported, the chemical manufacturer, importers, distributor, or employer shall add the information to the label before the chemical is shipped or introduced into the workplace again.

(g) **"Material safety data sheets."**

(1) Chemical manufacturers and importers shall obtain or develop a material safety data sheet for each hazardous chemical they produce or import. Employers shall have a material safety data sheet in the workplace for each hazardous chemical which they use.

(2) Each material safety data sheet shall be in English (although the employer may maintain copies in other languages as well), and shall contain at least the following information:

(i) The identity used on the label, and, except as provided for in paragraph (i) of this section on trade secrets:

(A) If the hazardous chemical is a single substance, its chemical and common name(s);

(B) If the hazardous chemical is a mixture which has been tested as a whole to determine its hazards, the chemical and common name(s) of the ingredients which contribute to these known hazards, and the common name(s) of the mixture itself; or,

(C) If the hazardous chemical is a mixture which has not been tested as a whole:

..1910.1200(g)(2)(i)(C)(1)

(1) The chemical and common name(s) of all ingredients which have been determined to be health hazards, and which comprise 1% or greater of the composition, except that chemicals identified as carcinogens under paragraph (d) of this section shall be listed if the concentrations are 0.1% or greater; and,

(2) The chemical and common name(s) of all ingredients which have been determined to be health hazards, and which comprise less than 1% (0.1% for carcinogens) of the mixture, if there is evidence that the ingredient(s) could be released from the mixture in concentrations which would exceed an established OSHA permissible exposure limit or ACGIH Threshold Limit Value, or could present a health risk to employees; and,

(3) The chemical and common name(s) of all ingredients which have been determined to present a physical hazard when present in the mixture;

(ii) Physical and chemical characteristics of the hazardous chemical (such as vapor pressure, flash point);

(iii) The physical hazards of the hazardous chemical, including the potential for fire, explosion, and reactivity;

(iv) The health hazards of the hazardous chemical, including signs and symptoms of exposure, and any medical conditions which are generally recognized as being aggravated by exposure to the chemical;

(v) The primary route(s) of entry;

..1910.1200(g)(2)(vi)

(vi) The OSHA permissible exposure limit, ACGIH Threshold Limit Value, and any other exposure limit used or recommended by the chemical manufacturer, importer, or employer preparing the material safety data sheet, where available;

(vii) Whether the hazardous chemical is listed in the National Toxicology Program (NTP) Annual Report on Carcinogens (latest edition) or has been found to be a potential carcinogen in the International Agency for Research on Cancer (IARC) Monographs (latest editions), or by OSHA;

(viii) Any generally applicable precautions for safe handling and use which are known to the chemical manufacturer, importer or employer preparing the material safety data sheet, including appropriate hygienic practices, protective measures during repair and maintenance of contaminated equipment, and procedures for clean-up of spills and leaks;

(ix) Any generally applicable control measures which are known to the chemical manufacturer, importer or employer preparing the material safety data sheet, such as appropriate engineering controls, work practices, or personal protective equipment;

(x) Emergency and first aid procedures;

(xi) The date of preparation of the material safety data sheet or the last change to it; and,

..1910.1200(g)(2)(xii)

(xii) The name, address and telephone number of the chemical manufacturer, importer, employer or other responsible party preparing or distributing the material safety data sheet, who can provide additional information on the hazardous chemical and appropriate emergency procedures, if necessary.

(3) If no relevant information is found for any given category on the material safety data sheet, the chemical manufacturer, importer or employer preparing the material safety data sheet shall mark it to indicate that no applicable information was found.

(4) Where complex mixtures have similar hazards and contents (i.e. the chemical ingredients are essentially the same, but the specific composition varies from mixture to mixture), the chemical manufacturer, importer or employer may prepare one material safety data sheet to apply to all of these similar mixtures.

(5) The chemical manufacturer, importer or employer preparing the material safety data sheet shall ensure that the information recorded accurately reflects the scientific evidence used in making the hazard determination. If the chemical manufacturer, importer or employer preparing the material safety data sheet becomes newly aware of any significant information regarding the hazards of a chemical, or ways to protect against the hazards, this new information shall be added to the material safety data sheet within three months. If the chemical is not currently being produced or imported the chemical manufacturer or importer shall add the information to the material safety data sheet before the chemical is introduced into the workplace again

...1910.1200(g)(6)

(6)

(i) Chemical manufacturers or importers shall ensure that distributors and employers are provided an appropriate material safety data sheet with their initial shipment, and with the first shipment after a material safety data sheet is updated;

(ii) The chemical manufacturer or importer shall either provide material safety data sheets with the shipped containers or send them to the distributor or employer prior to or at the time of the shipment;

(iii) If the material safety data sheet is not provided with a shipment that has been labeled as a hazardous chemical, the distributor or employer shall obtain one from the chemical manufacturer or importer as soon as possible; and,

(iv) The chemical manufacturer or importer shall also provide distributors or employers with a material safety data sheet upon request.

(7)

(i) Distributors shall ensure that material safety data sheets, and updated information, are provided to other distributors and employers with their initial shipment and with the first shipment after a material safety data sheet is updated;

(ii) The distributor shall either provide material safety data sheets with the shipped containers, or send them to the other distributor or employer prior to or at the time of the shipment;

..1910.1200(g)(7)(iii)

(iii) Retail distributors selling hazardous chemicals to employers having a commercial account shall provide a material safety data sheet to such employers upon request, and shall post a sign or otherwise inform them that a material safety data sheet is available;

(iv) Wholesale distributors selling hazardous chemicals to employers over-the-counter may also provide material safety data sheets upon the request of the employer at the time of the over-the-counter purchase, and shall post a sign or otherwise inform such employers that a material safety data sheet is available;

(v) If an employer without a commercial account purchases a hazardous chemical from a retail distributor not required to have material safety data sheets on file (i.e., the retail distributor does not have commercial accounts and does not use the materials), the retail distributor shall provide the employer, upon request, with the name, address, and telephone number of the chemical manufacturer, importer, or distributor from which a material safety data sheet can be obtained;

(vi) Wholesale distributors shall also provide material safety data sheets to employers or other distributors upon request; and,

(vii) Chemical manufacturers, importers, and distributors need not provide material safety data sheets to retail distributors that have informed them that the retail distributor does not sell the product to commercial accounts or open the sealed container to use it in their own workplaces.

..1910.1200(g)(8)

(8) The employer shall maintain in the workplace copies of the required material safety data sheets for each hazardous chemical, and shall ensure that they are readily accessible during each work shift to employees when they are in their work area(s). (Electronic access, microfiche, and other alternatives to maintaining paper copies of the material safety data sheets are permitted as long as no barriers to immediate employee access in each workplace are created by such options.)

(9) Where employees must travel between workplaces during a workshift, i.e., their work is carried out at more than one geographical location, the material safety data sheets may be kept at the primary workplace facility. In this situation, the employer shall ensure that employees can immediately obtain the required information in an emergency.

(10) Material safety data sheets may be kept in any form, including operating procedures, and may be designed to cover groups of hazardous chemicals in a work area where it may be more appropriate to address the hazards of a process rather than individual hazardous chemicals. However, the employer shall ensure that in all cases the required information is provided for each hazardous chemical, and is readily accessible during each work shift to employees when they are in their work area(s).

(11) Material safety data sheets shall also be made readily available, upon request, to designated representatives and to the Assistant Secretary, in accordance with the requirements of 29 CFR 1910.1020(e). The Director shall also be given access to material safety data sheets in the same manner.

..1910.1200(h)

(h) **"Employee information and training."**

(1) Employers shall provide employees with effective information and training on hazardous chemicals in their work area at the time of their initial assignment, and whenever a new physical or health hazard the employees have not previously been trained about is introduced into their work area. Information and training may be designed to cover categories of hazards (e.g., flammability, carcinogenicity) or specific chemicals. Chemical-specific information must always be available through labels and material safety data sheets.

(2) "Information." Employees shall be informed of:

(i) The requirements of this section;

(ii) Any operations in their work area where hazardous chemicals are present; and,

(iii) The location and availability of the written hazard communication program, including the required list(s) of hazardous chemicals, and material safety data sheets required by this section.

(3) "Training." Employee training shall include at least:

(i) Methods and observations that may be used to detect the presence or release of a hazardous chemical in the work area (such as monitoring conducted by the employer, continuous monitoring devices, visual appearance or odor of hazardous chemicals when being released, etc.);

(ii) The physical and health hazards of the chemicals in the work area;

..1910.1200(h)(3)(iii)

(iii) The measures employees can take to protect themselves from these hazards, including specific procedures the employer has implemented to protect employees from exposure to hazardous chemicals, such as appropriate work practices, emergency procedures, and personal protective equipment to be used; and,

(iv) The details of the hazard communication program developed by the employer, including an explanation of the labeling system and the material safety data sheet, and how employees can obtain and use the appropriate hazard information.

(i) **"Trade secrets."**

(1) The chemical manufacturer, importer, or employer may withhold the specific chemical identity, including the chemical name and other specific identification of a hazardous chemical, from the material safety data sheet, provided that:

(i) The claim that the information withheld is a trade secret can be supported;

(ii) Information contained in the material safety data sheet concerning the properties and effects of the hazardous chemical is disclosed;

(iii) The material safety data sheet indicates that the specific chemical identity is being withheld as a trade secret; and,

(iv) The specific chemical identity is made available to health professionals, employees, and designated representatives in accordance with the applicable provisions of this paragraph.

..1910.1200(i)(2)

(2) Where a treating physician or nurse determines that a medical emergency exists and the specific chemical identity of a hazardous chemical is necessary for emergency or first-aid treatment, the chemical manufacturer, importer, or employer shall immediately disclose the specific chemical identity of a trade secret chemical to that treating physician or nurse, regardless of the existence of a written statement of need or a confidentiality agreement. The chemical manufacturer, importer, or employer may require a written statement of need and confidentiality agreement, in accordance with the provisions of paragraphs (i)(3) and (4) of this section, as soon as circumstances permit.

(3) In non-emergency situations, a chemical manufacturer, importer, or employer shall, upon request, disclose a specific chemical identity, otherwise permitted to be withheld under paragraph (i)(1) of this section, to a health professional (i.e. physician, industrial hygienist, toxicologist, epidemiologist, or occupational health nurse) providing medical or other occupational health services to exposed employee(s), and to employees or designated representatives, if:

(i) The request is in writing;

(ii) The request describes with reasonable detail one or more of the following occupational health needs for the information:

(A) To assess the hazards of the chemicals to which employees will be exposed;

(B) To conduct or assess sampling of the workplace atmosphere to determine employee exposure levels;

(C) To conduct pre-assignment or periodic medical surveillance of exposed employees;

(D) To provide medical treatment to exposed employees;

..1910.1200(i)(3)(ii)(E)

(E) To select or assess appropriate personal protective equipment for exposed employees;

(F) To design or assess engineering controls or other protective measures for exposed employees; and,

(G) To conduct studies to determine the health effects of exposure.

(iii) The request explains in detail why the disclosure of the specific chemical identity is essential and that, in lieu thereof, the disclosure of the following information to the health professional, employee, or designated representative, would not satisfy the purposes described in paragraph (i)(3)(ii) of this section:

(A) The properties and effects of the chemical;

(B) Measures for controlling workers' exposure to the chemical;

(C) Methods of monitoring and analyzing worker exposure to the chemical; and,

(D) Methods of diagnosing and treating harmful exposures to the chemical;

(iv) The request includes a description of the procedures to be used to maintain the confidentiality of the disclosed information; and,

..1910.1200(i)(3)(v)

(v) The health professional, and the employer or contractor of the services of the health professional (i.e. downstream employer, labor organization, or individual employee), employee, or designated representative, agree in a written confidentiality agreement that the health professional, employee, or designated representative, will not use the trade secret information for any purpose other than the health need(s) asserted and agree not to release the information under any circumstances other than to OSHA, as provided in paragraph (i)(6) of this section, except as authorized by the terms of the agreement or by the chemical manufacturer, importer, or employer.

(4) The confidentiality agreement authorized by paragraph (i)(3)(iv) of this section:

(i) May restrict the use of the information to the health purposes indicated in the written statement of need;

(ii) May provide for appropriate legal remedies in the event of a breach of the agreement, including stipulation of a reasonable pre-estimate of likely damages; and,

(iii) May not include requirements for the posting of a penalty bond.

(5) Nothing in this standard is meant to preclude the parties from pursuing non-contractual remedies to the extent permitted by law.

(6) If the health professional, employee, or designated representative receiving the trade secret information decides that there is a need to disclose it to OSHA, the chemical manufacturer, importer, or employer who provided the information shall be informed by the health professional, employee, or designated representative prior to, or at the same time as, such disclosure.

..1910.1200(i)(7)

(7) If the chemical manufacturer, importer, or employer denies a written request for disclosure of a specific chemical identity, the denial must:

(i) Be provided to the health professional, employee, or designated representative, within thirty days of the request;

(ii) Be in writing;

(iii) Include evidence to support the claim that the specific chemical identity is a trade secret;

(iv) State the specific reasons why the request is being denied; and,

(v) Explain in detail how alternative information may satisfy the specific medical or occupational health need without revealing the specific chemical identity.

(8) The health professional, employee, or designated representative whose request for information is denied under paragraph (i)(3) of this section may refer the request and the written denial of the request to OSHA for consideration.

(9) When a health professional, employee, or designated representative refers the denial to OSHA under paragraph (i)(8) of this section, OSHA shall consider the evidence to determine if:

..1910.1200(i)(9)(i)

(i) The chemical manufacturer, importer, or employer has supported the claim that the specific chemical identity is a trade secret;

(ii) The health professional, employee, or designated representative has supported the claim that there is a medical or occupational health need for the information; and,

(iii) The health professional, employee or designated representative has demonstrated adequate means to protect the confidentiality.

(10)

(i) If OSHA determines that the specific chemical identity requested under paragraph (i)(3) of this section is not a "bona fide" trade secret, or that it is a trade secret, but the requesting health professional, employee, or designated representative has a legitimate medical or occupational health need for the information, has executed a written confidentiality agreement, and has shown adequate means to protect the confidentiality of the information, the chemical manufacturer, importer, or employer will be subject to citation by OSHA.

..1910.1200(i)(10)(ii)

(ii) If a chemical manufacturer, importer, or employer demonstrates to OSHA that the execution of a confidentiality agreement would not provide sufficient protection against the potential harm from the unauthorized disclosure of a trade secret specific chemical identity, the Assistant Secretary may issue such orders or impose such additional limitations or conditions upon the disclosure of the requested chemical information as may be appropriate to assure that the occupational health services are provided without an undue risk of harm to the chemical manufacturer, importer, or employer.

(11) If a citation for a failure to release specific chemical identity information is contested by the chemical manufacturer, importer, or employer, the matter will be adjudicated before the Occupational Safety and Health Review Commission in accordance with the Act's enforcement scheme and the applicable Commission rules of procedure. In accordance with the Commission rules, when a chemical manufacturer, importer, or employer continues to withhold the information during the contest, the Administrative Law Judge may review the citation and supporting documentation "in camera" or issue appropriate orders to protect the confidentiality of such matters.

(12) Notwithstanding the existence of a trade secret claim, a chemical manufacturer, importer, or employer shall, upon request, disclose to the Assistant Secretary any information which this section

requires the chemical manufacturer, importer, or employer to make available. Where there is a trade secret claim, such claim shall be made no later than at the time the information is provided to the Assistant Secretary so that suitable determinations of trade secret status can be made and the necessary protections can be implemented.

(13) Nothing in this paragraph shall be construed as requiring the disclosure under any circumstances of process or percentage of mixture information which is a trade secret.

..1910.1200(j)

(j) **"Effective dates."**

Chemical manufacturers, importers, distributors, and employers shall be in compliance with all provisions of this section by March 11, 1994.Note: The effective date of the clarification that the exemption of wood and wood products from the Hazard Communication standard in paragraph (b)(6)(iv) only applies to wood and wood products including lumber which will not be processed, where the manufacturer or importer can establish that the only hazard they pose to employees is the potential for flammability or combustibility, and that the exemption does not apply to wood or wood products which have been treated with a hazardous chemical covered by this standard, and wood which may be subsequently sawed or cut generating dust has been stayed from March 11, 1994 to August 11, 1994.[59 FR 17479, April 13, 1994; 59 FR 65947, Dec. 22, 1994; 61 FR 5507, Feb. 13, 1996]

OSHA Regulations (Standards - 29 CFR) - Table of Contents

# Appendix 3

## OSHA Regulations (Standards – 29 CFR) Respiratory Protection – 1910.134

**Standard Number: 1910.134**
Standard Title: Respiratory Protection
Subpart Number: I
Subpart Title: Personal Protective Equipment

Please refer to Standard CFR 1910.134.
NOTICE: Effective April 8, 1998, this standard will be changing its designation from 1910.134 to 1910.139 and will apply to TB. However, employers may continue to follow the old 1910.134 (1997) until the start-up date of the new standard October 5, 1998.

---

Sec. 1910.134 Respiratory protection.

This section applies to General Industry (part 1910), Shipyards (part 1915), Marine Terminals (part 1917), Longshoring (part 1918), and Construction (part 1926).

(a) **Permissible practice.**

(1) In the control of those occupational diseases caused by breathing air contaminated with harmful dusts, fogs, fumes, mists, gases, smokes, sprays, or vapors, the primary objective shall be to prevent atmospheric contamination. This shall be accomplished as far as feasible by accepted engineering control measures (for example, enclosure or confinement of the operation, general and local ventilation, and substitution of less toxic materials). When effective engineering controls are not feasible, or while they are being instituted, appropriate respirators shall be used pursuant to this section.

(2) Respirators shall be provided by the employer when such equipment is necessary to protect the health of the employee. The employer shall provide the respirators which are applicable and suitable for

the purpose intended. The employer shall be responsible for the establishment and maintenance of a respiratory protection program which shall include the requirements outlined in paragraph (c) of this section.

(b) **Definitions.**

The following definitions are important terms used in the respiratory protection standard in this section.

Air-purifying respirator means a respirator with an air-purifying filter, cartridge, or canister that removes specific air contaminants by passing ambient air through the air-purifying element.

Assigned protection factor (APF) [Reserved]

Atmosphere-supplying respirator means a respirator that supplies the respirator user with breathing air from a source independent of the ambient atmosphere, and includes supplied-air respirators (SARs) and self-contained breathing apparatus (SCBA) units.

Canister or cartridge means a container with a filter, sorbent, or catalyst, or combination of these items, which removes specific contaminants from the air passed through the container.

Demand respirator means an atmosphere-supplying respirator that admits breathing air to the facepiece only when a negative pressure is created inside the facepiece by inhalation.

Emergency situation means any occurrence such as, but not limited to, equipment failure, rupture of containers, or failure of control equipment that may or does result in an uncontrolled significant release of an airborne contaminant.

Employee exposure means exposure to a concentration of an airborne contaminant that would occur if the employee were not using respiratory protection.

End-of-service-life indicator (ESLI) means a system that warns the respirator user of the approach of the end of adequate respiratory protection, for example, that the sorbent is approaching saturation or is no longer effective.

Escape-only respirator means a respirator intended to be used only for emergency exit.

Filter or air purifying element means a component used in respirators to remove solid or liquid aerosols from the inspired air.

Filtering facepiece (dust mask) means a negative pressure particulate respirator with a filter as an integral part of the facepiece or with the entire facepiece composed of the filtering medium.

Fit factor means a quantitative estimate of the fit of a particular respirator to a specific individual, and typically estimates the ratio of the concentration of a substance in ambient air to its concentration inside the respirator when worn.

Fit test means the use of a protocol to qualitatively or quantitatively evaluate the fit of a respirator on an individual. (See also Qualitative fit test QLFT and Quantitative fit test QNFT.)

Helmet means a rigid respiratory inlet covering that also provides head protection against impact and penetration.

High efficiency particulate air (HEPA) filter means a filter that is at least 99.97% efficient in removing monodisperse particles of 0.3 micrometers in diameter. The equivalent NIOSH 42 CFR 84 particulate filters are the N100, R100, and P100 filters.

Hood means a respiratory inlet covering that completely covers the head and neck and may also cover portions of the shoulders and torso.

Immediately dangerous to life or health (IDLH) means an atmosphere that poses an immediate threat to life, would cause irreversible adverse health effects, or would impair an individual's ability to escape from a dangerous atmosphere.

Interior structural firefighting means the physical activity of fire suppression, rescue or both, inside of buildings or enclosed structures which are involved in a fire situation beyond the incipient stage. (See 29 CFR 1910.155)

Loose-fitting facepiece means a respiratory inlet covering that is designed to form a partial seal with the face.

Maximum use concentration (MUC) [Reserved].

Negative pressure respirator (tight fitting) means a respirator in which the air pressure inside the facepiece is negative during inhalation with respect to the ambient air pressure outside the respirator.

Oxygen deficient atmosphere means an atmosphere with an oxygen content below 19.5% by volume.

Physician or other licensed health care professional (PLHCP) means an individual whose legally permitted scope of practice (i.e., license, registration, or certification) allows him or her to independently provide, or be delegated the responsibility to provide, some or all of the health care services required by paragraph (e) of this section.

Positive pressure respirator means a respirator in which the pressure inside the respiratory inlet covering exceeds the ambient air pressure outside the respirator.

Powered air-purifying respirator (PAPR) means an air-purifying respirator that uses a blower to force the ambient air through air-purifying elements to the inlet covering.

Pressure demand respirator means a positive pressure atmosphere-supplying respirator that admits breathing air to the facepiece when the positive pressure

[[Page 1271]]

is reduced inside the facepiece by inhalation.

Qualitative fit test (QLFT) means a pass/fail fit test to assess the adequacy of respirator fit that relies on the individual's response to the test agent.

Quantitative fit test (QNFT) means an assessment of the adequacy of respirator fit by numerically measuring the amount of leakage into the respirator.

Respiratory inlet covering means that portion of a respirator that forms the protective barrier between the user's respiratory tract and an air-purifying device or breathing air source, or both. It may be a facepiece, helmet, hood, suit, or a mouthpiece respirator with nose clamp.

Self-contained breathing apparatus (SCBA) means an atmosphere-supplying respirator for which the breathing air source is designed to be carried by the user.

Service life means the period of time that a respirator, filter or sorbent, or other respiratory equipment provides adequate protection to the wearer.

Supplied-air respirator (SAR) or airline respirator means an atmosphere-supplying respirator for which the source of breathing air is not designed to be carried by the user.

This section means this respiratory protection standard.

Tight-fitting facepiece means a respiratory inlet covering that forms a complete seal with the face.

User seal check means an action conducted by the respirator user to determine if the respirator is properly seated to the face.

(c) **Respiratory protection program.**

This paragraph requires the employer to develop and implement a written respiratory protection program with required worksite-specific procedures and elements for required respirator use. The program must be administered by a suitably trained program administrator. In addition, certain program elements may be required for voluntary use to prevent potential hazards associated with the use of the respirator. The Small Entity Compliance Guide contains criteria for the selection of a program administrator and a sample program that meets the requirements of this paragraph. Copies of the Small Entity Compliance Guide will be available on or about April 8, 1998 from the Occupational Safety and Health Administration's Office of Publications, Room N 3101, 200 Constitution Avenue, NW, Washington, DC, 20210 (202-219-4667).

(1) In any workplace where respirators are necessary to protect the health of the employee or whenever respirators are required by the employer, the employer shall establish and implement a written respiratory protection program with worksite-specific procedures. The program shall be updated as necessary to reflect those changes in workplace conditions that affect respirator use. The employer shall include in the program the following provisions of this section, as applicable:

(i) Procedures for selecting respirators for use in the workplace;

(ii) Medical evaluations of employees required to use respirators;

(iii) Fit testing procedures for tight-fitting respirators;

(iv) Procedures for proper use of respirators in routine and reasonably foreseeable emergency situations;

(v) Procedures and schedules for cleaning, disinfecting, storing, inspecting, repairing, discarding, and otherwise maintaining respirators;

(vi) Procedures to ensure adequate air quality, quantity, and flow of breathing air for atmosphere-supplying respirators;

(vii) Training of employees in the respiratory hazards to which they are potentially exposed during routine and emergency situations;

(viii) Training of employees in the proper use of respirators, including putting on and removing them, any limitations on their use, and their maintenance; and

(ix) Procedures for regularly evaluating the effectiveness of the program.

(2) Where respirator use is not required:

(i) An employer may provide respirators at the request of employees or permit employees to use their own respirators, if the employer determines that such respirator use will not in itself create a hazard. If the employer determines that any voluntary respirator use is permissible, the employer shall provide the respirator users with the information contained in Appendix D to this section ("Information for Employees Using Respirators When Not Required Under the Standard"); and

(ii) In addition, the employer must establish and implement those elements of a written respiratory protection program necessary to ensure that any employee using a respirator voluntarily is medically able to use that respirator, and that the respirator is cleaned, stored, and maintained so that its use does not present a health hazard to the user. Exception: Employers are not required to include in a written respiratory protection program those employees whose only use of respirators involves the voluntary use of filtering facepieces (dust masks).

(3) The employer shall designate a program administrator who is qualified by appropriate training or experience that is commensurate with the complexity of the program to administer or oversee the respiratory protection program and conduct the required evaluations of program effectiveness.

(4) The employer shall provide respirators, training, and medical evaluations at no cost to the employee.

(d) **Selection of respirators.**

This paragraph requires the employer to evaluate respiratory hazard(s) in the workplace, identify relevant workplace and user factors, and base respirator selection on these factors. The paragraph also specifies appropriately protective respirators for use in IDLH atmospheres, and limits the selection and use of air-purifying respirators.

(1) General requirements. (i) The employer shall select and provide an appropriate respirator based on the respiratory hazard(s) to which the worker is exposed and workplace and user factors that affect respirator performance and reliability.

(ii) The employer shall select a NIOSH-certified respirator. The respirator shall be used in compliance with the conditions of its certification.

(iii) The employer shall identify and evaluate the respiratory hazard(s) in the workplace; this evaluation shall include a reasonable estimate of employee exposures to respiratory hazard(s) and an identification of the contaminant's chemical state and physical form. Where the employer cannot identify or reasonably estimate the employee exposure, the employer shall consider the atmosphere to be IDLH.

(iv) The employer shall select respirators from a sufficient number of respirator models and sizes so that the respirator is acceptable to, and correctly fits, the user.

(2) Respirators for IDLH atmospheres.

(i) The employer shall provide the following respirators for employee use in IDLH atmospheres:

(A) A full facepiece pressure demand SCBA certified by NIOSH for a minimum service life of thirty minutes, or

(B) A combination full facepiece pressure demand supplied-air respirator (SAR) with auxiliary self-contained air supply.

(ii) Respirators provided only for escape from IDLH atmospheres shall be NIOSH-certified for escape from the atmosphere in which they will be used.

(iii) All oxygen-deficient atmospheres shall be considered IDLH. Exception: If the employer demonstrates that, under all foreseeable conditions, the oxygen concentration can be maintained within

[[Page 1272]]

the ranges specified in Table II of this section (i.e., for the altitudes set out in the table), then any atmosphere-supplying respirator may be used.

(3) Respirators for atmospheres that are not IDLH.

(i) The employer shall provide a respirator that is adequate to protect the health of the employee and ensure compliance with all other OSHA statutory and regulatory requirements, under routine and reasonably foreseeable emergency situations.

(A) Assigned Protection Factors (APFs) [Reserved]

(B) Maximum Use Concentration (MUC) [Reserved]

(ii) The respirator selected shall be appropriate for the chemical state and physical form of the contaminant.

(iii) For protection against gases and vapors, the employer shall provide:

(A) An atmosphere-supplying respirator, or

(B) An air-purifying respirator, provided that:

(1) The respirator is equipped with an end-of-service-life indicator (ESLI) certified by NIOSH for the contaminant; or

(2) If there is no ESLI appropriate for conditions in the employer's workplace, the employer implements a change schedule for canisters and cartridges that is based on objective information or data that will ensure that canisters and cartridges are changed before the end of their service life. The employer shall describe in the respirator program the information and data relied upon and the basis for the canister and cartridge change schedule and the basis for reliance on the data.

(iv) For protection against particulates, the employer shall provide:

(A) An atmosphere-supplying respirator; or

(B) An air-purifying respirator equipped with a filter certified by NIOSH under 30 CFR part 11 as a high efficiency particulate air (HEPA) filter, or an air-purifying respirator equipped with a filter certified for particulates by NIOSH under 42 CFR part 84; or

(C) For contaminants consisting primarily of particles with mass median aerodynamic diameters (MMAD) of at least 2 micrometers, an air-purifying respirator equipped with any filter certified for particulates by NIOSH.

Table I.—Assigned Protection Factors [Reserved]

Table II

| Altitude (ft.) | Oxygen deficient Atmospheres (% $O_2$) for which the employer may rely on atmosphere-supplying respirators |
|---|---|
| Less than 3,001 | 16.0-19.5 |
| 3,001-4,000 | 16.4-19.5 |
| 4,001-5,000 | 17.1-19.5 |
| 5,001-6,000 | 17.8-19.5 |
| 6,001-7,000 | 18.5-19.5 |
| 7,001-8,000\1\ | 19.3-19.5 |

\1\ Above 8,000 feet the exception does not apply. Oxygen-enriched breathing air must be supplied above 14,000 feet.

(e) **Medical evaluation.**

Using a respirator may place a physiological burden on employees that varies with the type of respirator worn, the job and workplace conditions in which the respirator is used, and the medical status of the employee. Accordingly, this paragraph specifies the minimum requirements for medical evaluation that employers must implement to determine the employee's ability to use a respirator.

(1) General. The employer shall provide a medical evaluation to determine the employee's ability to use a respirator, before the employee is fit tested or required to use the respirator in the workplace. The employer may discontinue an employee's medical evaluations when the employee is no longer required to use a respirator.

(2) Medical evaluation procedures.

(i) The employer shall identify a physician or other licensed health care professional (PLHCP) to perform medical evaluations using a medical questionnaire or an initial medical examination that obtains the same information as the medical questionnaire.

(ii) The medical evaluation shall obtain the information requested by the questionnaire in Sections 1 and 2, Part A of Appendix C of this section.

(3) Follow-up medical examination.

(i) The employer shall ensure that a follow-up medical examination is provided for an employee who gives a positive response to any question among questions 1 through 8 in Section 2, Part A of Appendix C or whose initial medical examination demonstrates the need for a follow-up medical examination.

(ii) The follow-up medical examination shall include any medical tests, consultations, or diagnostic procedures that the PLHCP deems necessary to make a final determination.

(4) Administration of the medical questionnaire and examinations.

(i) The medical questionnaire and examinations shall be administered confidentially during the employee's normal working hours or at a time and place convenient to the employee. The medical questionnaire shall be administered in a manner that ensures that the employee understands its content.

(ii) The employer shall provide the employee with an opportunity to discuss the questionnaire and examination results with the PLHCP.

(5) Supplemental information for the PLHCP.

(i) The following information must be provided to the PLHCP before the PLHCP makes a recommendation concerning an employee's ability to use a respirator:

(A) The type and weight of the respirator to be used by the employee;

(B) The duration and frequency of respirator use (including use for rescue and escape);

(C) The expected physical work effort;

(D) Additional protective clothing and equipment to be worn; and

(E) Temperature and humidity extremes that may be encountered.

(ii) Any supplemental information provided previously to the PLHCP regarding an employee need not be provided for a subsequent medical evaluation if the information and the PLHCP remain the same.

(iii) The employer shall provide the PLHCP with a copy of the written respiratory protection program and a copy of this section.

Note to Paragraph (e)(5)(iii): When the employer replaces a PLHCP, the employer must ensure that the new PLHCP obtains this information, either by providing the documents directly to the PLHCP or having the documents transferred from the former PLHCP to the new PLHCP. However, OSHA does not expect employers to have employees medically reevaluated solely because a new PLHCP has been selected.

(6) Medical determination. In determining the employee's ability to use a respirator, the employer shall:

(i) Obtain a written recommendation regarding the employee's ability to use the respirator from the PLHCP. The recommendation shall provide only the following information:

(A) Any limitations on respirator use related to the medical condition of the employee, or relating to the workplace conditions in which the respirator will be used, including whether or not the employee is medically able to use the respirator;

(B) The need, if any, for follow-up medical evaluations; and

(C) A statement that the PLHCP has provided the employee with a copy of the PLHCP's written recommendation.

[[Page 1273]]

(ii) If the respirator is a negative pressure respirator and the PLHCP finds a medical condition that may place the employee's health at increased risk if the respirator is used, the employer shall provide a PAPR if the PLHCP's medical evaluation finds that the employee can use such a respirator; if a subsequent medical evaluation finds that the employee is medically able to use a negative pressure respirator, then the employer is no longer required to provide a PAPR.

(7) Additional medical evaluations. At a minimum, the employer shall provide additional medical evaluations that comply with the requirements of this section if:

(i) An employee reports medical signs or symptoms that are related to ability to use a respirator;

(ii) A PLHCP, supervisor, or the respirator program administrator informs the employer that an employee needs to be reevaluated;

(iii) Information from the respiratory protection program, including observations made during fit testing and program evaluation, indicates a need for employee reevaluation; or

(iv) A change occurs in workplace conditions (e.g., physical work effort, protective clothing, temperature) that may result in a substantial increase in the physiological burden placed on an employee.

(f) **Fit testing.**

This paragraph requires that, before an employee may be required to use any respirator with a negative or positive pressure tight-fitting facepiece, the employee must be fit tested with the same make, model, style, and size of respirator that will be used. This paragraph specifies the kinds of fit tests allowed, the procedures for conducting them, and how the results of the fit tests must be used.

(1) The employer shall ensure that employees using a tight-fitting facepiece respirator pass an appropriate qualitative fit test (QLFT) or quantitative fit test (QNFT) as stated in this paragraph.

(2) The employer shall ensure that an employee using a tight-fitting facepiece respirator is fit tested prior to initial use of the respirator, whenever a different respirator facepiece (size, style, model or make) is used, and at least annually thereafter.

(3) The employer shall conduct an additional fit test whenever the employee reports, or the employer,

PLHCP, supervisor, or program administrator makes visual observations of, changes in the employee's physical condition that could affect respirator fit. Such conditions include, but are not limited to, facial scarring, dental changes, cosmetic surgery, or an obvious change in body weight.

(4) If after passing a QLFT or QNFT, the employee subsequently notifies the employer, program administrator, supervisor, or PLHCP that the fit of the respirator is unacceptable, the employee shall be given a reasonable opportunity to select a different respirator facepiece and to be retested.

(5) The fit test shall be administered using an OSHA-accepted QLFT or QNFT protocol. The OSHA-accepted QLFT and QNFT protocols and procedures are contained in Appendix A of this section.

(6) QLFT may only be used to fit test negative pressure air-purifying respirators that must achieve a fit factor of 100 or less.

(7) If the fit factor, as determined through an OSHA-accepted QNFT protocol, is equal to or greater than 100 for tight-fitting half facepieces, or equal to or greater than 500 for tight-fitting full facepieces, the QNFT has been passed with that respirator.

(8) Fit testing of tight-fitting atmosphere-supplying respirators and tight-fitting powered air-purifying respirators shall be accomplished by performing quantitative or qualitative fit testing in the negative pressure mode, regardless of the mode of operation (negative or positive pressure) that is used for respiratory protection.

# Appendix 4

## MATERIAL SAFETY DATA SHEET

**Stoddard Solvent**
**MSDS Number: S6588**
**Effective Date: 12/08/96**

### 1. Product Identification

**Synonyms:** White spirits; Mineral spirits type I; Petroleum distillate
**CAS No.:** 8052-41-3
**Molecular Weight:** Not applicable to mixtures
**Chemical Formula:** > 65% $C_{10}$ or higher hydrocarbons
**Product Codes:** V110

### 2. Composition/Information on Ingredients

| Ingredient | CAS No | Percent | Hazardous |
|---|---|---|---|
| Stoddard Solvent | 8052-41-3 | 98 - 100% | Yes |

### 3. Hazards Identification

**Emergency Overview**

DANGER! HARMFUL OR FATAL IF SWALLOWED. AFFECTS CENTRAL NERVOUS SYSTEM. MAY AFFECT KIDNEYS. FLAMMABLE LIQUID AND VAPOR. HARMFUL IF INHALED. CAUSES IRRITATION TO SKIN, EYES AND RESPIRATORY TRACT.

**MATERIAL SAFETY DATA SHEET**

**Stoddard Solvent**
**MSDS Number: S6588**
**Effective Date: 12/08/96**

**3. Hazards Identification (cont.)**

**Safety Ratings** (Provided here for your convenience)

| | |
|---|---|
| **Health Rating:** | 1 – Slight |
| **Flammability Rating:** | 2 – Moderate |
| **Reactivity Rating:** | 0 – None |
| **Contact Rating:** | 1 – Slight |
| **Lab Protective Equip:** | GOGGLES; LAB COAT; VENT HOOD; PROPER GLOVES; CLASS |
| **Storage Color Code:** | Red (Flammable) |

**Potential Health Effects**

| | |
|---|---|
| **Inhalation:** | Effects are typically those of most hydrocarbons, dizziness and euphoria leading to unconsciousness in severe cases. Vapors also irritate the respiratory tract. Symptoms may include coughing, difficult breathing and chest pain. A central nervous system depressant. |
| **Ingestion:** | Fatal dose for humans estimated at 3-4 oz, but ingestion of much smaller amounts may cause lung edema and possible death because of aspiration into lungs. |
| **Skin Contact:** | The defatting action of this solvent may lead to soreness, inflammation and, possibly, dermatitis. |
| **Eye Contact:** | Vapors may be irritating at concentrations of 450 ppm and above (15 minutes exposure) and contact with the liquid solvent can be painful and possibly damaging to eye tissues. |
| **Chronic Exposure:** | Chronic exposure may lead to central nervous system complications, blood changes (aplastic anemia, a rare occurrence that is potentially fatal), and dermatitis. Animal studies have indicated the potential for liver and kidney damage. |
| **Aggravation of Pre-existing Conditions:** | Persons with pre-existing skin disorders or eye problems or impaired kidney function may be more susceptible to the effects of the substance. |

**4. First Aid Measures**

| | |
|---|---|
| **Inhalation:** | Remove to fresh air. If not breathing, give artificial respiration. If breathing is difficult, give oxygen. Get medical attention. |
| **Ingestion:** | Aspiration hazard. If swallowed, vomiting may occur spontaneously, but DO NOT INDUCE. If vomiting occurs, keep head below hips to prevent aspiration into lungs. Never give anything by mouth to an unconscious person. Call a physician immediately. |
| **Skin Contact:** | Immediately flush skin with plenty of soap and water for at least 15 minutes. Remove contaminated clothing and shoes. Get medical attention. Wash clothing before reuse. Thoroughly clean shoes before reuse. |
| **Eye Contact:** | Immediately flush eyes with plenty of water for at least 15 minutes, lifting lower and upper eyelids occasionally. Get medical attention immediately. |

**5. Fire Fighting Measures**

| | | |
|---|---|---|
| **Fire:** | Flash point: | 38ºC (100ºF) |
| | CC Autoignition temperature: | 232 – 260ºC (450 – 500ºF) |
| | Flammable limits in air % by volume: | lel: 0.8; uel: ca. 6 Flammable |
| | This liquid is near its lower flammability limit at room temperature. Flash point may range between 38-40ºC. Contact with strong oxidizers may cause fire. | |
| **Explosion:** | Above flash point, vapor-air mixtures are explosive within flammable limits noted above. Sealed containers may rupture when heated. Sensitive to static discharge. | |
| **Fire Extinguishing Media:** | Water spray, dry chemical, alcohol foam, or carbon dioxide. Direct stream of water can scatter and spread flames. | |
| **Special Information:** | In the event of a fire, wear full protective clothing and NIOSH-approved self-contained breathing apparatus with full facepiece operated in the pressure demand or other positive pressure mode. Water spray may be used to keep fire-exposed containers cool. | |

**MATERIAL SAFETY DATA SHEET**

**Stoddard Solvent**
**MSDS Number: S6588**
**Effective Date: 12/08/96**

**6. Accidental Release Measures**

Ventilate area of leak or spill. Remove all sources of ignition. Wear appropriate personal protective equipment as specified in Section 8. Isolate hazard area. Keep unnecessary and unprotected personnel from entering. Contain and recover liquid when possible. Use non-sparking tools and equipment. Collect liquid in an appropriate container or absorb with an inert material (e. g., vermiculite, dry sand, earth), and place in a chemical waste container. Do not use combustible materials, such as sawdust. Do not flush to sewer! If a leak or spill has not ignited, use water spray to disperse the vapors, to protect personnel attempting to stop leak, and to flush spills away from exposures.

**7. Handling and Storage**

Protect against physical damage. Store in a cool, dry well-ventilated location, away from any area where the fire hazard may be acute. Outside or detached storage is preferred. Separate from incompatibles. Containers should be bonded and grounded for transfers to avoid static sparks. Storage and use areas should be No Smoking areas. Use non-sparking type tools and equipment, including explosion proof ventilation. Containers of this material may be hazardous when empty since they retain product residues (vapors, liquid); observe all warnings and precautions listed for the product. Do Not attempt to clean empty containers since residue is difficult to remove. Do not pressurize, cut, weld, braze, solder, drill, grind or expose such containers to heat, sparks, flame, static electricity or other sources of ignition: they may explode and cause injury or death.

**8. Exposure Controls/Personal Protection**

**Airborne Exposure Limits:** OSHA Permissible Exposure Limit (PEL): 500 ppm (TWA) -ACGIH Threshold Limit Value (TLV): 100 ppm (TWA)

**Ventilation System:** A system of local and/or general exhaust is recommended to keep employee exposures below the Airborne Exposure Limits. Local exhaust ventilation is generally preferred because it can control the emissions of the contaminant at its source, preventing dispersion of it into the general work area. Please refer to the ACGIH document, *Industrial Ventilation, A Manual of Recommended Practices*, most recent edition, for details. Use explosion-proof equipment.

**Personal Respirators (NIOSH Approved):** If the exposure limit is exceeded, a half-face organic vapor respirator may be worn for up to ten times the exposure limit or the maximum use concentration specified by the appropriate regulatory agency or respirator supplier, whichever is lowest. A full-face piece organic vapor respirator may be worn up to 50 times the exposure limit or the maximum use concentration specified by the appropriate regulatory agency or respirator supplier, whichever is lowest. For emergencies or instances where the exposure levels are not known, use a full-face piece positive-pressure, air-supplied respirator. WARNING: Air-purifying respirators do not protect workers in oxygen-deficient atmospheres.

**Skin Protection:** Rubber or neoprene gloves and additional protection including impervious boots, apron, or coveralls, as needed in areas of unusual exposure.

**Eye Protection:** Use chemical safety goggles and/or a full face shield where splashing is possible. Maintain eye wash fountain and quick-drench facilities in work area.

**9. Physical and Chemical Properties**

| | |
|---|---|
| **Appearance:** | Clear, colorless liquid |
| **Odor:** | Kerosene-like |
| **Solubility:** | Insoluble in water |
| **Density:** | 0.79 |
| **pH:** | No information found |
| **% Volatiles by volume @ 21°C (70°F):** | 100 |
| **Boiling Point:** | 156 - 202°C (313 - 396°F) |
| **Melting Point:** | No information found |
| **Vapor Density (Air=1):** | 4.8 |
| **Vapor Pressure (mm Hg):** | ca. 5 @ 25°C (77°F) |
| **Evaporation Rate (BuAc=1):** | ca. 0.08 |

**MATERIAL SAFETY DATA SHEET**

**Stoddard Solvent**
**MSDS Number: S6588**
**Effective Date: 12/08/96**

**10. Stability and Reactivity**

**Stability:** Stable under ordinary conditions of use and storage.

**Hazardous Decomposition Products:** Carbon dioxide and carbon monoxide may form when heated to decomposition.

**Hazardous Polymerization:** Will not occur.

**Incompatibilities:** Strong acids, strong oxidizers.

**Conditions to Avoid:** Heat, flames, ignition sources and incompatibles.

**11 Toxicological Information**

No $LD_{50}/LC_{50}$ information found relating to normal routes of occupational exposure. Investigated as a tumorigen.

**Cancer Lists**

| —NTP Carcinogen— | | | |
|---|---|---|---|
| **Ingredient** | **Known** | **Anticipated** | **IARC Category** |
| Stoddard Solvent (8052-41-3) | No | No | None |

**12. Ecological Information**

**Environmental Fate:** No information found.
**Environmental Toxicity:** No information found.

**13. Disposal Considerations**

Whatever cannot be saved for recovery or recycling should be handled as hazardous waste and sent to a RCRA approved incinerator or disposed in a RCRA approved waste facility. Processing, use or contamination of this product may change the waste management options. State and local disposal regulations may differ from federal disposal regulations. Dispose of container and unused contents in accordance with federal, state and local requirements.

**14. Transport Information**

**Domestic (Land, D.O.T.)**

**Proper Shipping Name:** PETROLEUM DISTILLATES, N.O.S. (STODDARD SOLVENT)
**Hazard Class:** 3
**UN/NA:** UN1268
**Packing Group:** III
**Info. reported for product/size:** 20L

**International (Water, I.M.O.)**

**Proper Shipping Name:** PETROLEUM DISTILLATES, N.O.S. (STODDARD SOLVENT)
**Hazard Class:** 3
**UN/NA:** UN1268
**Packing Group:** III
**Info. reported for product/size:** 20L

**MATERIAL SAFETY DATA SHEET**

**Stoddard Solvent**
**MSDS Number: S6588**
**Effective Date: 12/08/96**

### 15. Regulatory Information

**Chemical Inventory Status – Part 1**

| Ingredient | TSCA | EC | Japan | Australia |
|---|---|---|---|---|
| Stoddard Solvent (8052-41-3) | Yes | Yes | No | Yes |

**Chemical Inventory Status – Part 2**

| | | —Canada— | | |
|---|---|---|---|---|
| Ingredient | Korea | DSL | NDSL | Phil. |
| Stoddard Solvent (8052-41-3) | Yes | Yes | No | Yes |

**Federal, State & International Regulations – Part 1**

| | —SARA 302— | | —SARA 313— | |
|---|---|---|---|---|
| Ingredient | RQ | TPQ | List | Chemical Catg. |
| Stoddard Solvent (8052-41-3) | No | No | No | No |

**Federal, State & International Regulations – Part 2**

| | —RCRA— | —TSCA— | |
|---|---|---|---|
| Ingredient | CERCLA | 26133 | 8(d) |
| Stoddard Solvent (8052-41-3) | No | No | No |

**Chemical Weapons Convention:** No    TSCA 12(b): No    CDTA: No
SARA 311/312:    Acute: Yes    Chronic: Yes    Fire: Yes    Pressure: No
Reactivity: No    (Pure/Liquid)

**Australian Hazchem Code:** 3[Y]E
Poison Schedule: S5

**WHMIS:** This MSDS has been prepared according to the hazard criteria of the Controlled Products Regulations (CPR) and the MSDS contains all of the information required by the CPR.

### 16. Other Information

**NFPA Ratings:** Health: 3    Flammability: 2    Reactivity: 0

**Label Hazard Warning:** DANGER! HARMFUL OR FATAL IF SWALLOWED. AFFECTS CENTRAL NERVOUS SYSTEM MAY AFFECT KIDNEYS. FLAMMABLE LIQUID AND VAPOR. HARMFUL IF INHALED. CAUSES IRRITATION TO SKIN, EYES AND RESPIRATORY TRACT.

**Label Precautions:** Keep away from heat, sparks and flame. Avoid breathing vapor. Keep container closed. Use only with adequate ventilation. Wash thoroughly after handling. Avoid contact with eyes, skin and clothing.

## MATERIAL SAFETY DATA SHEET

**Stoddard Solvent**
**MSDS Number: S6588**
**Effective Date: 12/08/96**

**16. Other Information (cont.)**

**Label First Aid:** Aspiration hazard. If swallowed, vomiting may occur spontaneously, but DO NOT INDUCE. If vomiting occurs, keep head below hips to prevent aspiration into lungs. Never give anything by mouth to an unconscious person. Call a physician immediately. If inhaled, remove to fresh air. If not breathing, give artificial respiration. If breathing is difficult, give oxygen. Get medical attention. In case of contact, immediately flush eyes or skin with plenty of water for at least 15 minutes. Remove contaminated clothing and shoes. Wash clothing before reuse. Get medical attention.

**Product Use:** Laboratory Reagent.

**Revision Information:** Pure. New 16 section MSDS format, all sections have been revised.

**Disclaimer:**

The manufacturer provides the information contained herein in good faith but makes no representation as to its comprehensiveness or accuracy. This document is intended only as a guide to the appropriate precautionary handling of the material by a properly trained person using this product. Individuals receiving the information must exercise their independent judgment in determining its appropriateness for a particular purpose.

THE MANUFACTURER MAKES NO REPRESENTATIONS OR WARRANTIES, EITHER EXPRESS OR IMPLIED, INCLUDING WITHOUT LIMITATION ANY WARRANTIES OF MERCHANTABILITY, FITNESS FOR A PARTICULAR PURPOSE WITH RESPECT TO THE INFORMATION SET FORTH HEREIN OR THE PRODUCT TO WHICH THE INFORMATION REFERS. ACCORDINGLY, THE MANUFACTURER WILL NOT BE RESPONSIBLE FOR DAMAGES RESULTING FROM USE OF OR RELIANCE UPON THIS INFORMATION.

# MATERIAL SAFETY DATA SHEET

**Styrene (Stabilized)**
**MSDS Number: S6986**
**Effective Date: 12/08/96**

## 1. Product Identification

**Synonyms:** Styrene Monomer; Vinylbenzene; Phenylethylene; Styrol; Cinnamene
**CAS No.:** 100-42-5
**Molecular Weight:** 104.15
**Chemical Formula:** $C_6H_5CH:CH_2$
**Product Codes:** V091

## 2. Composition/Information on Ingredients

| Ingredient | CAS No | Percent | Hazardous |
|---|---|---|---|
| Styrene | 100-42-5 | 90 – 100% | Yes |

## 3. Hazards Identification

**Emergency Overview**

DANGER! FLAMMABLE! CAUSES BURNS. HARMFUL IF SWALLOWED OR INHALED.
TARGET ORGAN(S): Central nervous system, respiratory system, eyes, skin.

**Safety Ratings** (Provided here for your convenience)

**Health Rating:** 2 – Moderate
**Flammability Rating:** 3 – Severe (Flammable)
**Reactivity Rating:** 2 – Moderate
**Contact Rating:** 3 – Severe (Corrosive)
**Lab Protective Equip:** GOGGLES & SHIELD; LAB COAT & APRON; VENT HOOD; PROPER GLOVES; CLASS B EXTINGUISHER
**Storage Color Code:** Red (Flammable)

**Potential Health Effects**

**Inhalation:** Excessive inhalation respiratory system, irritation, may cause pulmonary edema, narcosis.

**Ingestion:** Irritation and burns to mouth and stomach.

**Skin Contact:** Burns.

**Eye Contact:** Irritation, burns.

**Chronic Exposure:** Central nervous system depression.

**Aggravation of Pre-existing Conditions:** No information found.

## 4. First Aid Measures

**Inhalation:** If inhaled, remove to fresh air. If not breathing, give artificial respiration. If breathing is difficult, give oxygen. Prompt action is essential.

**Ingestion:** Induce vomiting immediately as directed by medical personnel. Never give anything by mouth to an unconscious person. Get medical attention.

**Skin Contact:** In case of contact, immediately flush skin with plenty of water for at least 15 minutes while removing contaminated clothing and shoes. Wash clothing before re-use.

**Eye Contact:** In case of eye contact, immediately flush with plenty of water for at least 15 minutes.

**MATERIAL SAFETY DATA SHEET**

**Styrene (Stabilized)**
**MSDS Number: S6986**
**Effective Date: 12/08/96**

5. **Fire Fighting Measures**

| | | |
|---|---|---|
| **Fire:** | Flash point: | 33°C (88°F) |
| | CC Autoignition temperature: | 490°C (914°F) |
| | Flammable limits in air % by volume: | lel: 1.1; uel: 6.1 |

**Explosion:** Vapors may flow along surfaces to distant ignition sources and flash back. Closed containers exposed to heat may explode. Contact with strong oxidizers may cause fire.

**Fire Extinguishing Media:** Use alcohol foam, dry chemical or carbon dioxide. (Water may be ineffective.)

**Special Information:** Firefighters should wear proper protective equipment and self-contained breathing apparatus with full facepiece operated in positive pressure mode. Move exposed containers from fire area if it can be done without risk. Use water to keep fire-exposed containers cool.

6. **Accidental Release Measures**
Wear self-contained breathing apparatus and full protective clothing. Shut off ignition sources; no flares, smoking or flames in area. Stop leak if you can do so without risk. Use water spray to reduce vapors. Take up with sand or other non-combustible absorbent material and place into container for later disposal. Flush area with water.

7. **Handling and Storage**
Bond and ground containers when transferring liquid. Store in light-resistant containers. Containers of this material may be hazardous when empty since they retain product residues (vapors, liquid); observe all warnings and precautions listed for the product.

8. **Exposure Controls/Personal Protection**

**Airborne Exposure Limits:** -OSHA Permissible Exposure Limit (PEL): 100 ppm (TWA) PEL (Ceiling) = 200 ppm.
-ACGIH Threshold Limit Value (TLV): 50 ppm (TWA), 100 ppm (STEL)

**Ventilation System:** A system of local and/or general exhaust is recommended to keep employee exposures below the Airborne Exposure Limits. Local exhaust ventilation is generally preferred because it can control the emissions of the contaminant at its source, preventing dispersion of it into the general work area. Please refer to the ACGIH document, *Industrial Ventilation, A Manual of Recommended Practices*, most recent edition, for details.

**Personal Respirators (NIOSH Approved):** For conditions of use where exposure to the substance is apparent, consult an industrial hygienist. For emergencies, or instances where the exposure levels are not known, use a full-facepiece positive-pressure, air-supplied respirator. WARNING: Air purifying respirators do not protect workers in oxygen-deficient atmospheres.

**Skin Protection:** Wear impervious protective clothing, including boots, gloves, lab coat, apron or coveralls, as appropriate, to prevent skin contact.

**Eye Protection:** Use chemical safety goggles and/or a full face shield where splashing is possible. Maintain eye wash fountain and quick-drench facilities in work area.

9. **Physical and Chemical Properties**

| | |
|---|---|
| **Appearance:** | Prisms |
| **Odor:** | Odorless |
| **Solubility:** | Negligible (<0.1%) |
| **Specific Gravity:** | 0.91 |
| **pH:** | No information found |
| **% Volatiles by volume @ 21°C (70°F):** | 100 |
| **Boiling Point:** | 145°C (293°F) |
| **Melting Point:** | -31°C (-24°F) |
| **Vapor Density (Air=1):** | 3.6 |
| **Vapor Pressure (mm Hg):** | 5 @ 20°C (68°F) |
| **Evaporation Rate (BuAc=1):** | No information found |

## MATERIAL SAFETY DATA SHEET

**Styrene (Stabilized)**
**MSDS Number: S6986**
**Effective Date: 12/08/96**

### 10. Stability and Reactivity

**Stability:** Stable under ordinary conditions of use and storage.

**Hazardous Decomposition Products:** Carbon monoxide, carbon dioxide.

**Hazardous Polymerization:** Will not occur.

**Incompatibilities:** Strong oxidizing agents, copper, strong acids, metallic salts, polymerization catalysts & accelerators.

**Conditions to Avoid:** Light, Heat, flame, other sources of ignition, air.

### 11 Toxicological Information

**Cancer Lists**

| —NTP Carcinogen— | | | |
|---|---|---|---|
| **Ingredient** | **Known** | **Anticipated** | **IARC Category** |
| Styrene (100-42-5) | No | No | 2B |

### 12. Ecological Information

**Environmental Fate:** No information found.
**Environmental Toxicity:** No information found.

### 13. Disposal Considerations

Whatever cannot be saved for recovery or recycling should be handled as hazardous waste and sent to a RCRA approved waste facility. Processing, use or contamination of this product may change the waste management options. State and local disposal regulations may differ from federal disposal regulations. Dispose of container and unused contents in accordance with federal, state and local requirements.

### 14. Transport Information

**Domestic (Land, D.O.T.)**

**Proper Shipping Name:** STYRENE MONOMER, INHIBITED
**Hazard Class:** 3
**UN/NA:** UN2055
**Packing Group:** III
**Info. reported for product/size:** 20L

**International (Water, I.M.O.)**

**Proper Shipping Name:** STYRENE MONOMER, INHIBITED
**Hazard Class:** 3.3
**UN/NA:** UN2055
**Packing Group:** III
**Info. reported for product/size:** 20L

**MATERIAL SAFETY DATA SHEET**

**Styrene (Stabilized)**
**MSDS Number: S6986**
**Effective Date: 12/08/96**

**14. Transport Information (cont.)**

**International (Water, I.M.O.)**

**Proper Shipping Name:** STYRENE MONOMER, INHIBITED
**Hazard Class:** 3.3
**UN/NA:** UN2055
**Packing Group:** III
**Info. reported for product/size:** 20L

**15. Regulatory Information**

**Chemical Inventory Status – Part 1**

| Ingredient | TSCA | EC | Japan | Australia |
|---|---|---|---|---|
| Styrene (100-42-5) | Yes | Yes | Yes | Yes |

**Chemical Inventory Status – Part 2**

| | | —Canada— | | |
|---|---|---|---|---|
| Ingredient | Korea | DSL | NDSL | Phil. |
| Styrene (100-42-5) | Yes | Yes | No | Yes |

**Federal, State & International Regulations – Part 1**

| | —SARA 302— | | —SARA 313— | |
|---|---|---|---|---|
| Ingredient | RQ | TPQ | List | Chemical Catg. |
| Styrene (100-42-5) | No | No | Yes | No |

**Federal, State & International Regulations – Part 2**

| | —RCRA— | —TSCA— | |
|---|---|---|---|
| Ingredient | CERCLA | 261.33 | 8(d) |
| Styrene (100-42-5) | 1000 | No | No |

**Chemical Weapons Convention:** No TSCA 12(b): No CDTA: No
SARA 311/312: Acute: Yes Chronic: Yes Fire: Yes Pressure: No
Reactivity: No (Pure/Liquid)

**Australian Hazchem Code:** 3[Y]
Poison Schedule: No information found.

**WHMIS:** This MSDS has been prepared according to the hazard criteria of the Controlled Products Regulations (CPR) and the MSDS contains all of the information required by the CPR.

**MATERIAL SAFETY DATA SHEET**

**Styrene (Stabilized)**
**MSDS Number: S6986**
**Effective Date: 12/08/96**

**16. Other Information**

**NFPA Ratings:** Health: 2 Flammability: 3 Reactivity: 2

**Label Hazard Warning:** DANGER! FLAMMABLE! CAUSES BURNS. HARMFUL IF SWALLOWED OR INHALED. TARGET ORGAN(S): Central nervous system, respiratory system, eyes, skin.

**Label Precautions:** Keep away from heat, sparks, flame. Do not get in eyes, on skin, on clothing. Avoid breathing vapor. Keep in tightly closed container. Use with adequate ventilation. Wash thoroughly after handling. In case of fire, use alcohol foam, dry chemical, carbon dioxide – water may be ineffective. Flush spill area with water spray. Prevent run-off from entering drains, sewers, or streams.

**Label First Aid:** If swallowed, induce vomiting immediately as directed by medical personnel. Never give anything by mouth to an unconscious person. If inhaled, remove to fresh air. If not breathing, give artificial respiration. If breathing is difficult, give oxygen. Prompt action is essential. In case of contact, immediately flush eyes or skin with plenty of water for at least 15 minutes while removing contaminated clothing and shoes. Wash clothing before reuse.

**Product Use:** Laboratory Reagent.

**Revision Information:** Pure. New 16 section MSDS format, all sections have been revised.

**Disclaimer:**

The manufacturer provides the information contained herein in good faith but makes no representation as to its comprehensiveness or accuracy. This document is intended only as a guide to the appropriate precautionary handling of the material by a properly trained person using this product. Individuals receiving the information must exercise their independent judgment in determining its appropriateness for a particular purpose.

THE MANUFACTURER MAKES NO REPRESENTATIONS OR WARRANTIES, EITHER EXPRESS OR IMPLIED, INCLUDING WITHOUT LIMITATION ANY WARRANTIES OF MERCHANTABILITY, FITNESS FOR A PARTICULAR PURPOSE WITH RESPECT TO THE INFORMATION SET FORTH HEREIN OR THE PRODUCT TO WHICH THE INFORMATION REFERS. ACCORDINGLY, THE MANUFACTURER WILL NOT BE RESPONSIBLE FOR DAMAGES RESULTING FROM USE OF OR RELIANCE UPON THIS INFORMATION.

# Appendix 5

## Advanced Exercises

### Introduction

The following scenario is based on an EPA Case Study used for Risk and Decision Making workshops. The situation described is hypothetical, but it is intended to demonstrate the real-life complexities of working in the environmental technology field. A variety of related activities have been developed, based on your understanding of the information presented. These activities are not intended for entry-level students, but rather for students that are either experienced in the field of environmental technology or nearing completion of their course of study.

### Background

You are the environmental health and safety coordinator for Electrobotics, a company that manufactures parts for Multichrome personal computers. The company employs 150 workers at its production plant. Electrobotics has been proactive for several years in implementing pollution prevention strategies. One of the strategies is to substitute nonhazardous materials for hazardous materials whenever possible. However, you and your colleagues have not found a workable substitution for one particular chemical – dinitrochickenwire (DNC).

There are a number of groundwater monitoring wells on the plant property from which you periodically draw samples. For the past two years, lab analyses of these samples have indicated that dinitrochickenwire is contaminating the groundwater. Concentrations of the chemical have remained constant over those years. The source of the contamination is certain to be the equalization lagoon on the site (i.e., a pond constructed to hold the output from the plant's wastewater treatment operation).

You shared your concerns about the groundwater problem with the company owner, to whom you report directly. You both agreed it would be proper to hire a consultant who could conduct an exposure assessment for the entire facility and the surrounding neighborhood. Upon completion of the assessment, the consultant submitted a written report. You and the owner understand the technical language in the report, but the owner would like it put into words others in the plant and surrounding community will understand.

**DNC Exposure Evaluation Report**

**Some Principles of Exposure Evaluation**

1. The purpose of exposure evaluation is to identify the magnitude of human exposure to dinitrochickenwire (DNC), the frequency and duration of that exposure, and the routes by which humans are exposed. It may also be useful to identify the number of exposed people along with other characteristics of the exposed population (e.g., age, sex).

2. Exposure may be based on measurement of the amount of DNC in various media (air, water, etc.) and knowledge of the amount of human intake of these media per unit of time (usually per day) under different conditions of activity.

3. Some individuals may be exposed by contact with several media. It is important to consider total intake from all media in such situations.

4. Because only a limited number of samples of various media can be taken for measurement, the representativeness of measured values of environmental contaminants is always uncertain. If sampling is adequately placed, the degree to which data for a given medium are representative of that medium can usually be known.

5. Sometimes air and water concentrations of pollutants can be estimated by mathematical models. Although some of these models are known to be predictive in many cases, they are not thought to be reliable in all cases.

6. Standard average values and ranges for human intake of various media are available and are generally used, unless data on specific agents indicate that such values are inappropriate.

## Site Description

Electrobotics manufactures parts for Multichrome personal computers and operates a production facility. The facility employs 150 workers and includes a wastewater treatment plant, and an equalization lagoon (surface impoundment). Wastewater from the manufacturing plant operations contains high concentrations of dinitrochickenwire (DNC). The wastewater is periodically discharged to an equalization lagoon, which is located within 200 feet of the western facility boundary. There are no other potential sources of DNC releases in the surrounding area. Adjacent to this boundary is a residential area of 20 houses in which 80 individuals reside. Electrobotics built an 8-foot high chain-link fence on the property boundary, separating the residents from the facility grounds. The area between the Electrobotics facility and the threatened water supply wells to the east is undeveloped and contains few residences. There has been some discussion about building housing on the undeveloped property, but no definite plans exist at this time. The Smith River, as shown in Figure 1, is located east of the facility about 1,500 feet from the facility boundary. There are no drinking water intakes along the river, but some recreational fishing and swimming occurs.

The equalization lagoon maintains a regular flow to an on-site activated sludge wastewater treatment facility. The equalization lagoon is 10,000 square feet and is maintained at an average fluid depth of 10 feet. The wastewater from the treatment plant is discharged to the river in compliance with the facility's NPDES permit.

The equalization lagoon is underlain by a compacted natural clay liner with a hydraulic conductivity (a measure of the rate at which it transmits water) of $3 \times 10^{-8}$ cm/sec, which is designed to provide substantial containment of the wastewater. The uppermost geological formation beneath the site is composed of approximately 75 feet of stratified glacial outwash that is composed of sand and gravel with some silt. These unconsolidated sediments form a productive aquifer that is a source of potable water to the surrounding area. A well field is located approximately 1.5 miles to the east (down gradient) of the site and provides approximately 3 million gallons per day (mgd) of potable water to its 50,000 customers. The well field contains 40 separate wells, each pumping 75,000 gallons per day (pgd).

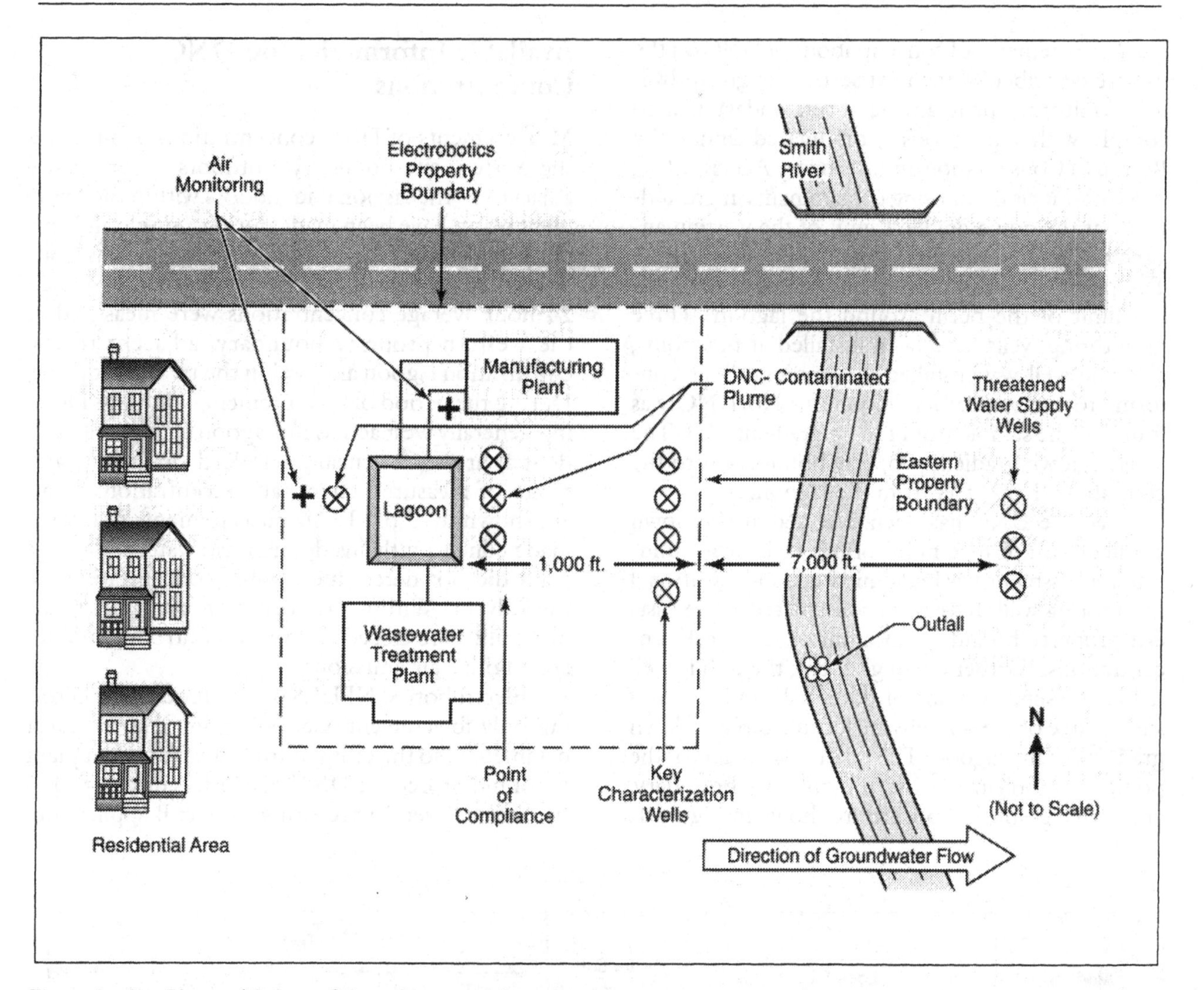


Figure 1: Site Plan and Points of Compliance and Exposure

The hydraulic conductivity of the aquifer has been estimated per day, based on field tests in the area. However, there may be a thin continuous layer of significantly lower or higher hydraulic conductivity within the glacial outwash. The water table is located about 10 feet below the bottom of the lagoon at the site. The hydraulic gradient (a measure of slope of the water table), based on the best available potentiometric head measurements, is estimated at 0.005 feet/foot toward the community well field.

The site is located in a humid area. The wind direction is seasonal; however, according to a wind rose from a local meteorological station (which describes wind direction and frequency), the wind blows west across the site toward the 20 homes adjacent to the lagoon approximately 30 percent of the time.

The owners of Electrobotics conducted an analysis of the lagoon in preparing their Part B application. The nature of the materials stored in the lagoon and the meteorological and hydrogeological conditions in the vicinity have resulted in some concern by EPA about possible exposure of workers at the facility and to residents in the vicinity to DNC via inhalation of air, as well as possible exposure of the community via contamination of drinking water supplies. In an attempt to respond to the regulations concerns of EPA staff, the owners undertook a program of air, surface water, and groundwater monitoring.

Measurements of concentrations of DNC on the site are described in the next section. A groundwater monitoring program has been undertaken to comply with regulations promulgated under the Resource Conservation and Recovery Act (RCRA). Electrobotics is monitoring contaminants in groundwater at the down gradient limit of the waste management unit, termed the "point of compliance." The point of compliance in this case is at the eastern limit of the berm around the lagoon. Three monitoring wells have been installed at this point, and one well was installed up gradient. This conforms to the regulatory minimum. No DNC was found in the sample from the up gradient well. The nearest down gradient property boundary is to the east, about 1,000 feet from the lagoon.

Because DNC has been detected at the down gradient compliance-monitoring wells, a groundwater assessment has been initiated, and additional monitoring wells have been constructed at the eastern property boundary to monitor chemical concentrations. Further down gradient, the public well field represents a point of potential human exposure. Three of the 40 wells are located directly down gradient of the lagoon. The relative locations of the point of compliance, eastern regulatory boundary, and actual point of exposure are shown in Figure 1.

## Available Information on DNC Concentrations

Measurements of DNC concentrations in air along the western site boundary, outdoors at the facility adjacent to the lagoon and indoors within the treatment plant, have been made during a single air-sampling program. The sampling program was conducted for a period of one week, during which 24-hour average concentrations were measured at the western property boundary, adjacent to the equalization lagoon and within the treatment plant. During the period of measurement, wind was blowing generally west across the lagoon toward the residential area. The mean, standard deviation, and range of measured chemical concentrations in air are shown in Table 1. Air measurements that were made concurrently inside the wastewater treatment plant did not detect measurable concentrations of the solvent. Within the treatment plant, all treatment units are closed and vapor-controlled to limit any fugitive air emissions.

Electrobotics' NPDES permit requires monitoring only for conventional pollutants and indicator parameters, so the effluent from the treatment plant is not monitored for DNC before it is discharged to the Smith River. In preparing its Part B application,

| Medium | Location | Detection Level | DNC | | |
|---|---|---|---|---|---|
| | | | Mean | Standard Deviation | Range |
| Air | Inside treatment plant | μg/m³ | ND[1] | — | — |
| Air | At western boundary of site | μg/m³ | 44 | 16 | 8–68 |
| Air | On-site | μg/m³ | 188 | 80 | 120–480 |
| Groundwater | Point of compliance | μg/liter | 332 | 30 | 290–350 |
| Groundwater | Eastern property boundary | μg/liter | BDL[2] | — | — |
| Groundwater | Public well field | μg/liter | ND[1] | | |
| Surface Water | Treatment plant fallout | μg/liter | ND[1] | — | — |

[1] Not detected at 1 μg/m³ for air or 1 μg/liter for water
[2] Trace concentrations below detection limit of 1 μg/liter

Table 1: Field Measurements of DNC Concentrations

| Location | DNC (µg/liter) |
|---|---|
| Eastern property boundary | 275 |
| Public well field | 5.5 |
| [1] Estimated time for DNC to travel from the lagoon to the property boundary is 1,000 days (about 3 years). The estimated travel time from the lagoon to the public well field is 8,000 days (about 22 years). | |

Table 2: Summary of Modeled Concentrations of DNC Steady State in Groundwater[1]

however, the owners of Electrobotics collected a limited number of samples from the treatment facility outfall. The results presented in Table 1 indicate that no DNC was detected.

The groundwater has been periodically sampled and tested for DNC both at the point of compliance, the monitoring wells on the eastern boundary, and at the water treatment plant at the community water system, which draws water from the aquifer for the municipal supply. Samples were taken at several depths. The concentrations of chemicals in monitoring wells at the point of compliance are indicated in Table 1.

The concentrations detected at the compliance point indicate contamination of the aquifer. Trace concentrations of DNC have also been detected at the eastern boundary in the most recent water tests, but at concentrations less than the method-detection limit of 1 µg/liter (or ppb). The municipal water supply has been tested, and no detectable concentrations of DNC were found. However, the detection limit for DNC offered by available analytical methods for groundwater is 1 µg/liter.

A very simple description of the area's geology and the status of the DNC-contaminated groundwater plume are presented in Figures 2 and 3. Available information suggests that the chemical plume

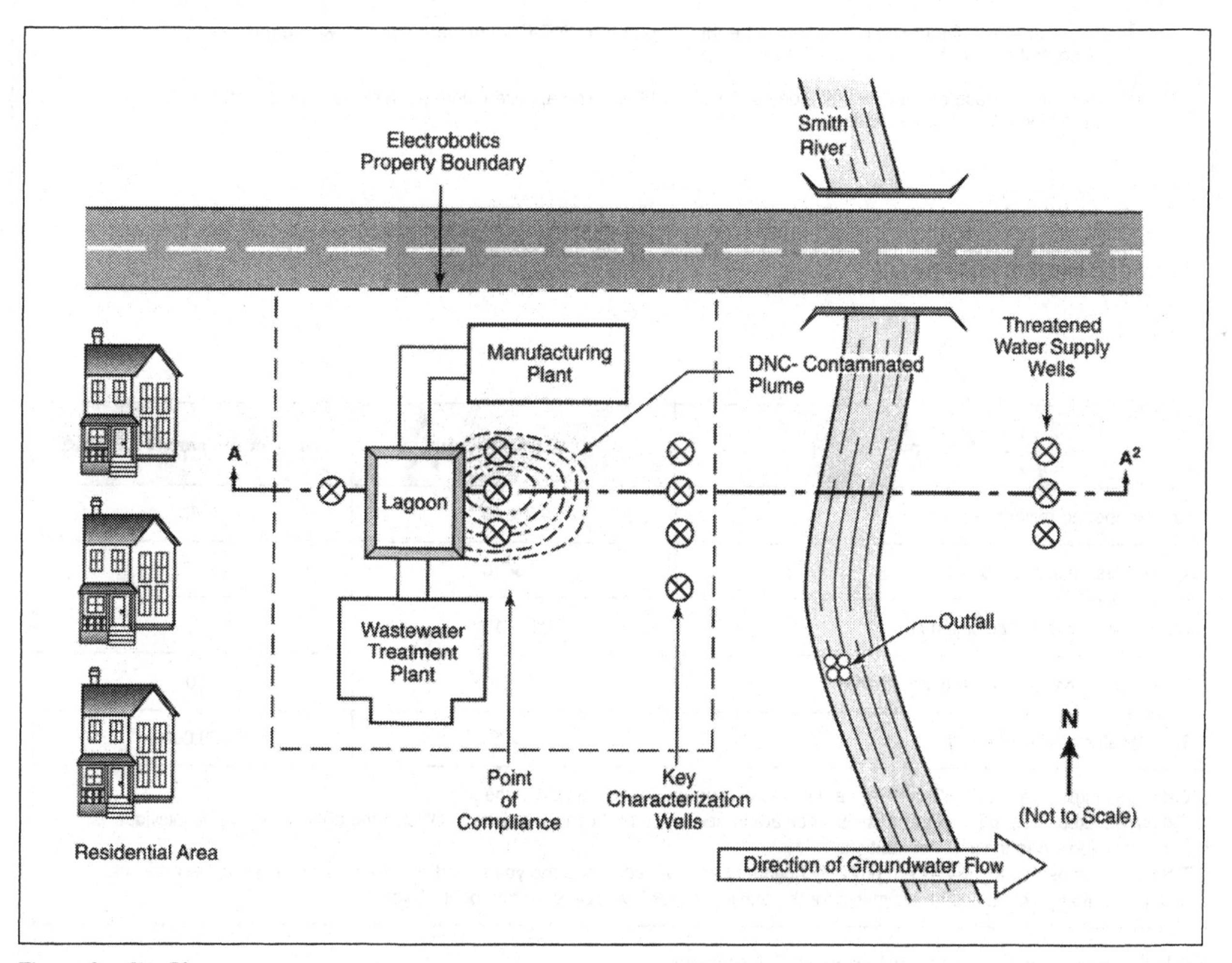


Figure 2: Site Plan

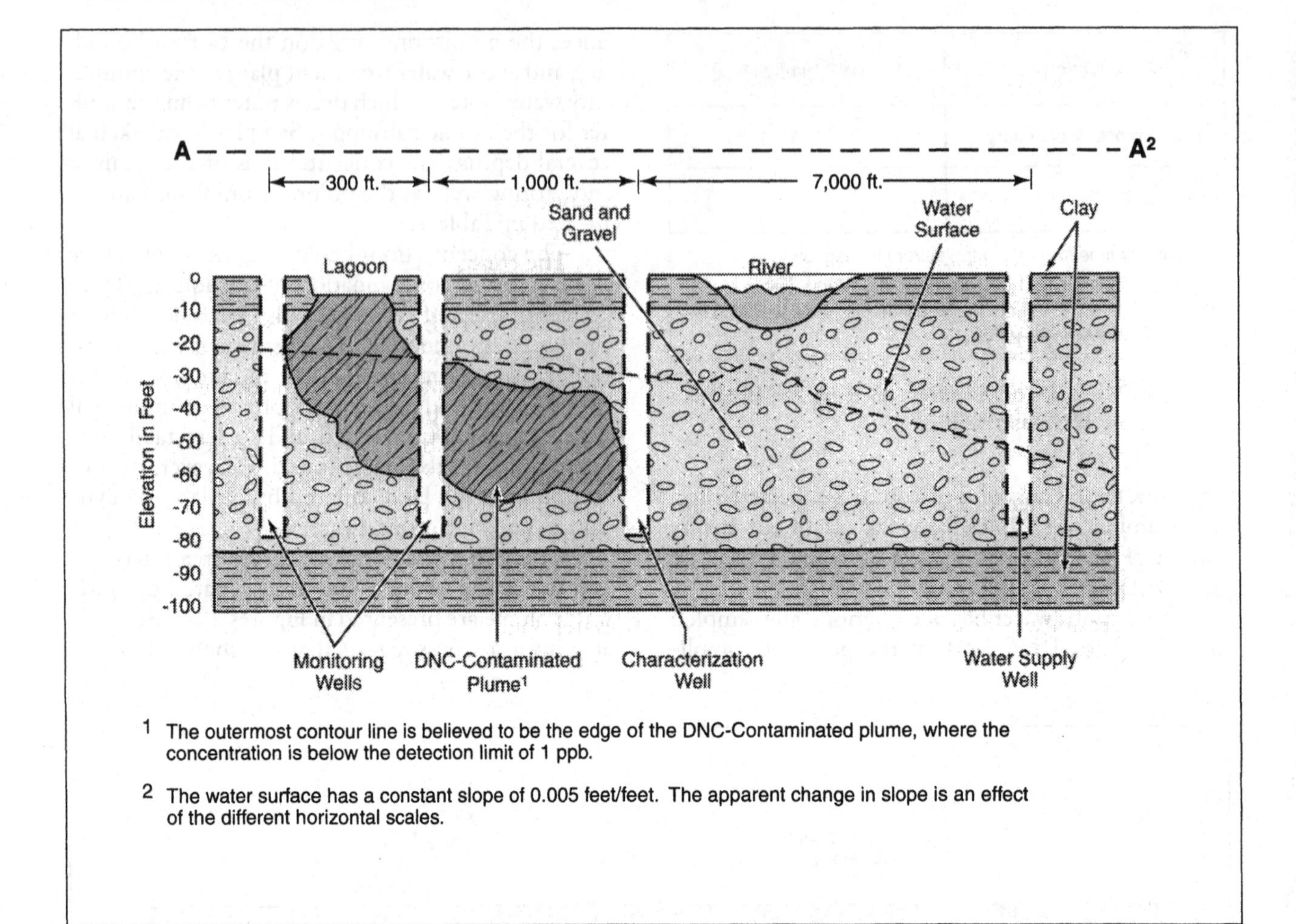


Figure 3: Geological cross section

| Medium | DNC (mg/kg/day) | Number of Persons Exposed |
|---|---|---|
| Air, neighboring residents[1] | $3.3 \times 10^{-3}$ | 80 |
| Air, workers on site (outdoors)[1] | $4.7 \times 10^{-3}$ | 150 |
| Groundwater, point of compliance[2] | $9.5 \times 10^{-3}$ | 0 |
| Groundwater, eastern property boundary[3] | $7.8 \times 10^{-3}$ | 0 |
| Groundwater, public well field[3] | $1.6 \times 10^{-4}$ | 50,000 |

[1]Estimated exposure from DNC in the air are based on current concentrations in the air.
[2]Estimated exposures at the point of compliance are based on current concentrations of DNC at the point, although no individual is currently exposed to those concentrations.
[3]Estimated exposures at the eastern property boundary will not occur for three years, and no one currently receives their drinking water from that point. Exposure estimates for the public well field will not occur for about 22 years.

Table 3: Summary of Results of Exposure Calculations

has not yet migrated beyond the eastern property boundary, but without corrective action, the concentrations are expected to increase in this vicinity over time, eventually affecting offsite groundwater.

A mathematical model of the contaminant plume movement in the aquifer has been constructed to estimate future concentrations in the aquifer. A summary of the modeled environmental concentrations in groundwater is presented in Table 2.

## Calculations of Human Exposure

The environmental concentrations summarized in Tables 1 and 2 are the starting point for the calculation of estimates of human exposure to DNC.

The medium in which a substance is present will determine the potential route of human exposure. For example, substances present in water may be ingested. Contaminated water may also lead to inhalation exposure when water is used for cooking or showering, although this exposure route is likely to be minor unless the contaminant is present at high concentrations in the water and is highly volatile. Exposure by inhalation of contaminated water or by the dermal route during bathing or showering was considered insignificant in this case because of the low concentrations of these substances in the municipal water supply and the fact that DNC is only moderately volatile. Doses resulting from these exposure routes were, therefore, not calculated. Substances present in air will be inhaled. Finally, substances present in soil may be ingested, absorbed through the skin, inhaled, or taken up by plants with human exposure resulting, if these plants are used as food or if these plants are fed to livestock, the various products of which are used as food.

Human exposure to contaminated soil, however, was considered insignificant at this site because soils potentially contaminated by groundwater are located at least 10 feet below the land surface. No concentrations of DNC were found in a limited set of soil samples taken from surface soils around the lagoon. Consequently, dose resulting from potential human exposure to soil was not calculated.

In order to estimate human exposure or dose of the constituent from each of the contaminated media, certain data and assumptions were applied. These data and assumptions relate to the extent and frequency of human contact with these media and the degree of absorption of chemicals for each route of exposure. When the constituent is ingested, a certain fraction will be absorbed through the gastrointestinal wall; when it is inhaled, a certain fraction will be absorbed by the lungs.

The method of dose calculation depends on the route of exposure. Certain standard values have been developed to estimate contact with and intake of certain media. For drinking water, the EPA and other scientific groups generally assume that adults drink 2 liters of water per day. It is also generally assumed that the average adult inhales from 20 to 23 cubic meters ($m^3$) of air per day. We have assumed an adult inhales 23 $m^3$/day, although EPA risk assessments are increasingly using 20 $m^3$/day. It should be noted, however, that the data and assumptions required for estimation of dose from most other routes of exposure are not as readily standardized.

The assumptions and calculations required to estimate DNC dose to the 80 neighboring residents resulting from inhalation of DNC-contaminated air are presented below to illustrate the elements of dose estimation.

## Inhalation of DNC-Contaminated Air by Neighboring Residents

**Assumptions:**

An adult inhales 23 $m^3$/day of air.

Based on wind direction analysis, the duration of exposure is 30 percent of the time on an annual average basis.

The body weight of an adult is 70 kg.

The inhalation absorption factor for DNC is 0.75.

The adult lives in the home throughout his lifetime.

**Calculations:**

$$\frac{0.044 \text{ mg(ave. DNC air conc. at boundary)}}{\text{m}^3} \times$$

$$\frac{23 \text{ m}^3}{\text{day}} \times \frac{1}{70 \text{ kg}} \times 0.3(\text{\% of time exposed}) \times$$

$$0.75 \text{ (inhalation absorption factor)} =$$

$$3.3 \times 10^{-3} \text{ mg/kg/day}$$

## Ecosystem Exposures

The ecosystem in the Smith River has been examined by a biologist employed by Electrobotics. The resulting study characterizes the area upstream and downstream from the facility. No major evidence of ecological damage was found, although there was some discoloration of the river around the outfall noted when the plant switched to the use of DNC. This discoloration has persisted. No threatened or endangered species were found near the Electrobotics site. However, several threatened or endangered species have been known to inhabit river ecosystems in the vicinity. Fish species include the Snail Darter, Slackwater Darter, Amber Darter, and Spotfin Chub. Terrestrial species include the Eastern Indigo Snake. Furthermore, several of these species are acutely sensitive to compounds similar in structure to DNC. The data available on DNC are not adequate for establishing ambient water quality criteria for aquatic life.

As noted earlier, the effluent discharged to the Smith River from the Electrobotics treatment plant was tested for DNC. No DNC was detected at a detection limit of 1 μg/liter.

## Remarks on Exposure Data

1. Ground water directly beneath and adjacent to the regulated unit has been routinely sampled and analyzed for the presence of DNC. Concentrations that have been measured have remained relatively constant for the past two years and indicate that DNC has leaked from the storage lagoon into the underlying aquifer. Groundwater moves toward a well field that provides drinking water for a community of 50,000 people. DNC has not been detected in the water supply, but may be present at concentrations below the current method detection limit, which is 1 μg/liter (ppb).

2. DNC is unstable in the environment and will decompose by biodegradation in shallow aquifer systems. The decomposition of the chemical in aquifers has been demonstrated in aquifer restoration programs at other sites, but the rates of decomposition are variable and have only been quantified in laboratory experiments. The rate of decomposition at concentrations less than 10 ppb is uncertain.

3. Trace concentrations (less than 1 μg/liter) of DNC have been detected at the eastern facility boundary. A mathematical model has been used to predict future concentrations of DNC in groundwater down gradient of the regulated unit. The model prediction indicates that the chemical will migrate in groundwater and degrade at a slow rate. Based on the model prediction, concentrations are expected to increase at the eastern property boundary if no corrective action is taken. The degradation products are not believed to be carcinogenic.

4. Estimates of future concentrations of DNC in drinking water are based on analytical predictions from a groundwater model. Appropriate adjustments regarding groundwater dilution were made to estimate concentrations at the tap.

5. Estimates of current exposure in air are based on analytical results obtained from measuring the air concentrations in the facility and at the boundary over a single seven-day period in April. Twenty-four hour composite samples were taken each day over the sampling period. No other air data are available.

6. Estimates of current concentrations of DNC in treatment plant effluent are based on analytical results from samples taken over the same seven-day period in April. Five samples were taken during this period.

# Activities

## Activity A

### Persuasive Letters and Memos

1. You believe that DNC is affecting the ecosystem around the Smith River. As a nearby property owner, write the company a letter and explain your reasons for wanting them to discontinue the use of DNC.
2. As the technician, explain in a memo to the owner why the exposure and risk to workers at the facility should not be considered in the same context as the exposure and risk to residents in the surrounding community.
3. As the technician, decide if Electrobotics has adequately characterized the nature of the groundwater contamination. Explain why you have taken the position. If your position is that the company has not adequately characterized the nature of the groundwater contamination, determine the additional types of information needed and explain why.
4. As a neighbor, you have some analytical data from the company regarding the exposure assessments. Explain in a letter to the owner your opinion of what is the most significant route of exposure to the members of the Smith River community. Provide reasons.

## Activity B

### Chain of Custody Form

You have collected a second sample of groundwater from monitoring well three because the well was inadvertently omitted during a recent sampling session. Fill out the chain of custody form(on the next page) that will accompany the sample to the laboratory.

—The sample number is EC-MW3-02-01.

—Create data such as phone number, Electrobotics address, date, your name (as collector), and laboratory name (for Sample Allocation).

—Use N/A for blanks that are not applicable (such as Waste Type Code).

—Explain the situation under Field Information.

—Sign your name, title, and date on line 1 of Chain of Possession

—Make up a name and title for a person at the lab who received your sample. Add the date (may be the same day you signed the form).

—Make up a name and title for the chemistry technician who analyzed your sample. Assume the sample was in that person's possession for multiple days.

Evaluate your form for legibility and completeness.

## Activity C

### Research a Concept

Select one of the concepts used in the consultant's written report listed below. Prepare a presentation to explain the concept to an audience that is unfamiliar with its exact meaning.

**Concepts:**

Ambient Water Quality Criteria

Biodegradation

Compliance Monitoring Well

Conventional Pollutants

Ecosystem

Hydraulic Conductivity

## Activity D

### Orally Interpret Data to the Local Citizens

Select any table or the calculation shown in the consultant's written report to Electrobotics. Prepare a presentation that explains the data to a local citizens group. Provide concrete comparisons of the data to help the listener envision the measurement. Your comparison can be either exact or approximate, for example 1 ppm is like a quarter cup of salt in a tanker truck of water. The goal is to help the citizens understand what the data means.

Collector's Sample No. ________

________

________

California Department of Health
Hazardous Materials Laborator

# CHAIN OF CUSTODY RECORD

## Hazardous Materials

Location of Sampling ________ Producer ________ Hauler

________ Disposal Site ________ Other

Company's Name ____________________ Telephone (   ) ________

Address ____________________
number street city state zip

Collector's Name ____________________ Telephone (   ) ________

Date Sampled ________ Time Sampled ________ hours

Type of Process Producing Waste ____________________

Waste Type Code ________ Other ________

Field Information ____________________

____________________

____________________

____________________

Sample Allocation:

1. ____________________
name of organization

2. ____________________
name of organization

3. ____________________
name of organization

| | signature | title | inclusive | date |
|---|---|---|---|---|
| 1. | | | | |
| 2. | | | | |
| 3. | | | | |

## Activity E

## Using Team Work to Revise a Document

The owner of Electrobotics does not think the Board of Directors will understand the consultant's written report. The Board members tend to have either a financial or marketing background. The strategy you decide to use is as follows:

1. Define terms
2. Develop an outline of the report
3. Write a revised report, using headings, bulleted lists, illustrations, and other formatting ideas

Form a three-member team to accomplish the three strategies listed above. Edit carefully so that you can submit a high quality document to the Board. An outline for the Site Description portion of the consultant's report is offered here to get you started on strategy 2. You may alter it if you wish. A possible list of terms from the report that may need defining is also provided.

I. Site Description
  A. Electrobotics
    1. Staff
    2. Product
    3. Facility
    4. Material of concern
  B. Within the plant boundaries
    1. Wastewater management
      a. Treatment plant
      b. Lagoon
    2. Monitoring/sampling targets
      a. Up gradient
      b. Down gradient
    3. Outside the plant boundaries
      a. Neighbors
      b. Well field
    4. Geology
      a. Stratified Glacial Outwash
      b. Aquifer
    5. Meteorology
      a. Humidity
      b. Wind Direction
    6. Sources of potential human exposure (identified by EPA)
      a. Air inhalation
      b. Drinking water supply
        1) Surface water
        2) Ground water
    7. Figure 1 Site Plan and Points of Compliance and Exposure

II. Available Information on DNC Concentrations

III. Calculations of Human Exposure

IV. Ecosystem Exposures

| Possible Terms Requiring Definition | |
|---|---|
| Activated Sludge | Media |
| Wastewater | Medium |
| Treatment Plant | Meteorological |
| Aquifer (Productive) | Microgram (µg) |
| Ambient Water Quality | MGD |
| Criteria | Municipal |
| Berm | NPDES Permit |
| Biodegradation | Outfall |
| Carcinogenic | Part B Application |
| Compliance Monitoring Well | PGD |
| Containment | Point of Compliance |
| Conventional Pollutants | Potable Water |
| Detection Level | Potentiometric Head Measurement |
| Detection Limit | Proactive |
| Dinitrochickenwire (DNC) | Promulgated |
| Dose | Quantified |
| Ecosystem | Range |
| Effluent | Regulatory Minimum |
| Equalization Lagoon | Representativeness |
| Fugitive Air Emissions | Resource Conservation and Recovery Act (RCRA) |
| Gastrointestinal | Route (of Exposure) |
| Gradient | Standard Deviation |
| Groundwater | Standardized |
| Hydraulic Conductivity | Steady State |
| Hydraulic Gradient | Stratified Glacial Outwash |
| Hydrogeological | Surface Impoundment |
| Indicator Parameters | Waste Management Unit |
| Ingest | Water Table |
| Inhalation Absorption | Well Field |
| Factor | |
| Inhale | |
| Mean | |

# Bibliography


## Chapter 1: Communication Skills Overview

Fast, Julius, *Body Language,* Pocket Books, 1988.

National Technical Information Service, *What Work Requires of Schools* (initial SCANS report). The report defines the five competencies and three-part foundation that constitute the SCANS skills. Single copies are available from National Technical Information Service (NTIS), Operations Division, Springfield, VA 22151, (703) 487-4650, NTIS number: PB92-146711 (call for current pricing).

**Internet Sites:**

Internet references may be accessed either through their http address or by typing the title of an article, enclosed in quotes, in the search box of a search engine.

Advanced Technology Environmental Education Center, http://www.ateec.org. This site contains many areas of interest to teachers and students.

Amazon.com, http://www.amazon.com is one example of an Internet bookseller that sells millions of titles. Use your favorite search engines to find additional booksellers.

National Archives and Records Administration, http://www.nara.gov/fedreg/ddhhome.html. "Federal Register Document Drafting Handbook," 1998 Edition, can be found on this page. The emphasis of the Handbook, which may be downloaded from the site, is on the use of plain English.

## Chapter 2: Writing and Technical Writing Basics

*The Chicago Manual of Style: The Essential Guide for Writers, Editors, and Publishers*, 14th Edition. Chicago, IL: University of Chicago Press, 1993.

Enger, Eldon D. and Bradley F. Smith, *Environmental Science: A Study of Interrelationships*, 6th Edition, New York, NY: WCB/McGraw-Hill, 1997. This environmental science textbook is written at a readable level, making it comprehensible for most students.

Hirsch, Donald. *Drafting Federal Law*, 2nd Edition. Washington D.C.: U.S. Government Printing Office, 1989.

HMTRI, *Living and Writing Responsibly: Environmental Readings for Classic and Technical Composition*. Raleigh, N.C.: HMTRI, 1992.

Leiner, Barry M., Vinton G. Cerf, David D. Clark, Robert E. Kahn, Leonard Kleinrock, Daniel C. Lynch, Jon Postel, Larry G. Roberts, Stephen Wolff, "A Brief History of the Internet," http://www.isoc.org/internet-history/

NIOSH/OSHA/USCG/EPA. *Occupational Safety and Health Guidance Manual for Hazardous Waste Site Activities*. Washington, DC: U.S. Government Printing Office, 1985. This publication is a helpful basic resource on these topics: Hazards, Planning and Organization, Training, Medical Program, Site Characterization, Air Monitoring, Personal Protective Equipment, Site Control, Decontamination, Handling Drums and Other Containers, and Site Emergencies.

Sindermann, Carl J., *Winning the Games Scientists Play*, New York, NY, Plenum Publishing Corp., 1982.

Strunk, William and E.B. White, *The Elements of Style*, 3rd Edition, Needham Heights, MA: Allyn & Bacon, 1995.

U.S. Government Printing Office, *The Federal Register: What It Is and How to Use It*. Washington, D.C.: U.S. Government Printing Office, 1992. This publication explains both the Federal Register and the Code of Federal Regulations.

Zinsser, William Knowlton, *On Writing Well: The Classic Guide to Writing Nonfiction*, 6th and up Edition, Harperreference, 1998.

**Internet Sites:**

Courtosi, Martin, University of Tennessee–Knoxville, "WWW Search Engines," http://www.lib.utk.edu/refs/search.html.

Gray, Terry A., Palomar College, "How to Search the Web: A Guide To Search Tools," http://daphne.palomar.edu/TGSEARCH/.

Purdue University Libraries, "Evaluation of Information Sources," http://thorplus.lib.purdue.edu/library_info/departments/ugrl/ref/bib/evalinfo.html.

Smith, Alastair, WWW Virtual Library, "WWW Virtual Library: Evaluation of Information Sources," http://www.vuw.ac.nz/~agsmith/evaln/evaln.htm.

## Chapter 3: Letters and Memos

On the Internet, find more information about general writing skills via "Writing," "Business Writing," "Technical Writing Skills," and "Proofreading and Editing Checklists."

Check amazon.com or other Internet book services to order handbooks on technical writing. In the title line, type "Technical Writing Handbook" or "Copyright Handbook" for a list of available titles.

Poe, Roy W. and Rosemary T. Fruehling, *Business Communication: A Problem-Solving Approach*, 4th Edition, New York, NY: McGraw-Hill, 1989.

Sabin, William A., *The Gregg Reference Manual*, 8th Edition. New York: Glencoe/MacMillan McGraw-Hill Book, 1996. Sabin's reference manual is widely used in business offices. The organization and format make it easy to find information about the conventions of spelling, punctuation, bibliographies, grammar and usage, and letter, memo, and report construction.

Strunk, William and E.B. White, *The Elements of Style*, 3rd Edition, Needham Heights, MA: Allyn &

Bacon, 1995. This tiny book (in the $6 range!) has long been considered a classic for writers who desire to tighten up their expression. Style recommendations such as avoiding beginning a sentence with "There is . . ." will help you write with impact.

U.S. Government Printing Office, *United States Government Printing Office Style Manual*, Washington, DC: Government Printing Office, 1984. If you work for a government agency such as EPA, you may be required to adhere to the conventions of GPO style when you prepare documents.

## Chapter 4: Technical Documents

HMTRI. *Living and Writing Responsibly: Environmental Readings for Classic and Technical Composition.* Raleigh: HMTRI, 1992.

**Internet Sites:**

The following Internet sites provide documents, such as material safety data sheets and agency regulations, upon which some of your technical environmental writing relies. You may also enter key words (MSDS or material safety data sheet) into a World Wide Web search engine.

Occupational Safety and Health Administration (OSHA), http://www.osha.gov

U.S. Department of Transportation (DOT), http://www.dot.gov

U.S. Environmental Protection Agency (EPA), http://www.epa.gov

## Chapter 5: Environmental Compliance Forms

Government forms may be found on the World Wide Web. Check the following agency URLs:

Occupational Safety and Health Administration (OSHA), http://www.osha.gov.

U.S. Department of Transportation, http://www.dot.gov.

U.S. Environmental Protection Agency, http://www.epa.gov.

Certification of Truth, Accuracy, and Completeness (Form CTAC), http://www.epa.gov/oar/oaqps/permits/comply.pdf.

Discharge Monitoring Report (DMR) (EPA Form 3320-1), http://www.epa.gov/earth1r6/6en/w/dmrf.pdf.

Emissions Unit Description for Fuel Combustion Sources (Form EUD-1), http://www.epa.gov/oar/oaqps/permits/unitdesc.pdf.

Emissions Unit Description for Process Sources (Form EUD-3), http://www.epa.gov/oar/oaqps/permits/unitdesc.pdf.

Emissions Unit Description for VOC Emitting Sources (Form EUD-2), http://www.epa.gov/oar/oaqps/permits/unitdesc.pdf.

Federal Operating Permit Application – General Information and Summary (Form GIS), http://www.epa.gov/oar/oaqps/permits/geninfo.pdf.

Hazardous Materials Incident Report (DOT Form F 5800.1), http://hazmat.dot.gov/5800.pdf.

Hazardous Waste Permit Application (EPA Form 8700-23), http://www.epa.gov/epaoswer/hazwaste/data/form8700/forms.htm.

Log and Summary of Occupational Injuries and Illnesses (OSHA No. 200), http://www.osha-slc.gov/Publications/osha200.pdf.

Material Safety Data Sheet (OSHA Form 174), http://www.osha-slc.gov/Publications/osha174.pdf.

Notice of Intent (NOI) for Storm Water Discharges Associated With Industrial Activity Under a NPDES Permit (EPA Form 3510-6), http://www.epa.gov/earth1r6/6en/w/sw/noiform.pdf.

Notification of Regulated Waste Activity (EPA Form 8700-12), http://www.epa.gov/epaoswer/hazwaste/data/form8700/forms.htm.

Pesticide Data Certification (EPA Form 8570-29), http://www.epa.gov/opprd001/forms/8570-29.pdf.

Pesticide Registration (EPA Form 8570-1), http://www.epa.gov/opprd001/forms/8570-1.pdf.

Pre-Manufacturing Notice (PMN) (EPA Form 7710-25), http://www.epa.gov/opptintr/newchms/pmnpart1.pdf and http://www.epa.gov/opptintr/newchms/pmnpart2.pdf.

Toxic Chemical Release Inventory, Form A (EPA Form 9350-2) Emergency Planning and Notification, Tier I/Tier II/TRI, http://www.epa.gov/opptintr/tri/toxdoc.pdf.

Toxic Chemical Release Inventory, Reporting Form R (EPA Form 9350-1), http://www.epa.gov/opptintr/tri/toxdoc.pdf.

Underground Storage Tank Notice (EPA Form 7530), http://www.epa.gov/swerust1/fedlaws/form7530.pdf.

Uniform Hazardous Waste Manifest (EPA Form 8700-22), http://www.epa.gov:80/epaoswer/hazwaste/gener/manifest/PDF/form.pdf.

## Chapter 6: Oral Communication

Briggs, Leslie J. ed., *Instructional Design: Principles and Applications,* 2nd Edition, Englewood Cliffs, NJ: Educational Technology Publications, 1991.

De Becker, Gavin, *The Gift of Fear*, Boston, MA: Little Brown and Company; 1997. De Becker, an expert on the prediction and management of violence, includes a chapter on managing workplace harassment that threatens to turn violent.

Hofstader, Robert and Kenneth Chapman, *Foundations for Excellence in the Chemical Process Industries, Voluntary Industry Standards for Chemical Process Industries Technical Workers.* Washington, DC: American Chemical Society, 1997.

McClure, Lynne Falkin, *Risky Business : Managing Violence in the Workplace*, Binghamton, NY: Haworth Press, 1996.

McVey, Steve and Thomas K. Capozzoli, *Managing Violence in the Workplace*, Boca Raton, FL: St. Lucie Press, 1996.

Schneid, Thomas D., *Occupational Health Guide to Violence in the Workplace*, Boca Raton, FL: Lewis Publishers, Inc, 1998.

**Internet Sites:**

American Chemical Society homepage, http://www.acs.org. A wealth of information of environmental interest.

Business Presenters Homepage, http://www.diversilink.com/bookstore/speaking.htm. Excellent one-stop site for books, courses, presentation tips and links to other Internet sources of information.

Equal Employment Opportunity Commission homepage, http://www.eeoc.gov.

Toastmasters, http://www.toastmasters.org. A nonprofit organization that provides the tools that enable people to develop or improve their skills in thinking, speaking, and listening. Contact for chapters near you and more information on the organization.

For additional Internet sources, try searching on "Presentation Skills" or "Public Speaking."

## Chapter 7: Coimmunication Skills Overview

Cook, John R., *The New Complete Guide to Environmental Careers*, 2nd Edition. Covelo, CA: Island Press, 1993.

Lain Kennedy, Joyce, *The Electronic Job Search Revolution*, 2nd Edition, New York, NY: John Wiley & Sons, Inc., 1995.

Molloy, John T., *John T. Molloy's New Dress for Success*, New York, NY: Warner Books, 1988.

Parker, Yana, *Blue Collar and Beyond: Résumés for Skilled Trades and Services.* Berkeley, CA: Ten Speed Press, 1995.

Parker, Yana, *The Damn Good Résumé Guide: A Crash Course in Résumé Writing,* 3rd Edition, Berkeley, CA: Ten Speed Press, 1996.

## Chapter 8: Interpersonal Skills


Belbin, Meredith, *Management Teams: Why They Succeed or Fail*, Woburn, MA: Butterworth-Heinemann, 1996.

Deming, W. Edwards, *Out of the Crisis*, Cambridge, MA: Massachusetts Institute of Technology, 1986.

Maddux, Robert, *Team Building: An Exercise in Leadership*, Menlo Park, CA: Crisp Publications, 1992.

Peters, Thomas and Robert H. Waterman, *In Search of Excellence: Lessons from American's Best-Run Companies*, New York, NY: Warner Books, 1988.

Peters, Tom, *Thriving on Chaos: Handbook for Management Revolution*, New York, NY: HarperCollins, 1991.

Walton, Mary, *The Deming Management Method*, New York, NY: Perigee Books, 1988.

Zemke, Ronald with Dick Schaaf, *The Service Edge*, New York, NY: Plume Publishing, 1990.


In the library, or on the Internet (amazon.com or another book seller) check for these authors' writings on organizational development and behavior/conflict management: Eric Berne, Rensus Likert, Gordon L. Lippitt, Tom Peters, Donald H. Weiss.

On the Internet, an interesting site about negotiation and mediation steps is PBS and Turner Adventure Learning's page: http://www.pbs.org/tal/un/mediation.html.

Another is Mental Health Net's "Psychological Self-Help" site about "I" Messages: http://www.cmhc.com/psyhelp/chap13/chap13g.htm.

# Index

## D



## E



## F



## G



## H



## I



## J



## K



## L



## M

## U



## V



## W